Large-Scale Genome Sequence Processing

Large-Scale Genome Sequence Processing

Masahiro Kasahara
Shinichi Morishita
University of Tokyo, Japan

Imperial College Press

Published by

Imperial College Press
57 Shelton Street
Covent Garden
London WC2H 9HE

Distributed by

World Scientific Publishing Co. Pte. Ltd.
5 Toh Tuck Link, Singapore 596224
USA office: 27 Warren Street, Suite 401-402, Hackensack, NJ 07601
UK office: 57 Shelton Street, Covent Garden, London WC2H 9HE

British Library Cataloguing-in-Publication Data
A catalogue record for this book is available from the British Library.

LARGE-SCALE GENOME SEQUENCE PROCESSING

Copyright © 2006 by Imperial College Press

All rights reserved. This book, or parts thereof, may not be reproduced in any form or by any means, electronic or mechanical, including photocopying, recording or any information storage and retrieval system now known or to be invented, without written permission from the Publisher.

For photocopying of material in this volume, please pay a copying fee through the Copyright Clearance Center, Inc., 222 Rosewood Drive, Danvers, MA 01923, USA. In this case permission to photocopy is not required from the publisher.

ISBN-13 978-1-86094-635-6
ISBN-10 1-86094-635-6

Printed in Singapore

To our parents

Preface

Efficient computer programs have made it possible to elucidate and analyze large-scale genomic sequences. Fundamental tasks, such as the assembly of numerous whole-genome shotgun fragments, the alignment of complementary DNA sequences with a long genome, and the design of gene-specific primers or oligomers, require efficient algorithms and state-of-the-art implementation techniques. This textbook emphasizes basic software implementation techniques for processing large-scale genome sequences.

In Chapters 1–6 we introduce the basic data structures and algorithms used for string matching and sequence alignment, describe performance-acceleration methods such as lookup tables and suffix arrays, and provide executable sample programs to motivate readers to try and enjoy programming. In Chapters 7 and 8 we show how these fundamental techniques are combined to solve major modern problems: whole-genome shotgun assembly, sequence alignment with a genome, comparative genomics, and the design of gene-specific sequence tags. These bioinformatics topics involve challenging, exciting, and real problems. Due to the space limitation, we cannot afford to include large executable programs in these two chapters.

This book is designed for use as a textbook during a semester course for advanced undergraduates and graduate students in computer science or biology. It is also suitable for computer scientists and programmers interested in genome sequence processing. To make the book accessible for nonprogrammers, there are many figures and examples to illustrate data structures and the behavior of algorithms.

We would like to thank Tomoyuki Yamada for studying seeded alignments with the second author, and Shin Sasaki for working with the first author to develop the whole genome shotgun assembler Ramen. This book owes a great deal to Tomoyuki and Shin.

Contents

Preface   vii

1. Simple String Search   1
   - 1.1 Storing a String in an Array 1
   - 1.2 Brute-Force String Search 2
   - 1.3 Encoding Strings into Integers 4
   - 1.4 Sorting k-mer Integers and a Binary Search 7
   - 1.5 Binary Search for the Boundaries of Blocks 8

2. Sorting   11
   - 2.1 Insertion Sort 11
   - 2.2 Merge Sort 12
   - 2.3 Worst-Case Time Complexity of Algorithms 16
   - 2.4 Heap Sort 17
   - 2.5 Randomized Quick Sort 22
   - 2.6 Improving the Performance of Quick Sort 27
   - 2.7 Ternary Split Quick Sort 32
   - 2.8 Radix Sort 33

3. Lookup Tables   39
   - 3.1 Direct-Address Tables 39
   - 3.2 Hash Tables 41
   - 3.3 Table Size 44
   - 3.4 Using the Frequencies of k-mers 45
   - 3.5 Techniques for Reducing Table Size 46

4. Suffix Arrays 51

4.1 Suffix Trees 52
4.2 Suffix Arrays 54
4.3 Binary Search of Suffix Arrays for Queries 56
4.4 Using the Longest Common Prefix Information to Accelerate the Search 58
4.5 Computing the Longest Common Prefixes 62
    4.5.1 Application to Occurrence Frequencies of k-mers . . 65
    4.5.2 Application to the Longest Common Factors 67
4.6 Suffix Array Construction – Doubling 68
4.7 Larsson-Sadakane Algorithm 69
4.8 Linear-Time Suffix Array Construction 74
4.9 A Note on Practical Performance 81

5. Space-Efficient String Search 83

5.1 Rabin-Karp Algorithm 83
5.2 Accelerating the Brute-Force String Search 86
5.3 Knuth-Morris-Pratt Algorithm 88
5.4 Bad Character Heuristics 94

6. Approximate String Search 99

6.1 Edit Operations and Alignments 101
6.2 Edit Graphs and Dynamic Programming 103
6.3 Needleman-Wunsch Algorithm 105
6.4 Smith-Waterman Algorithm for Computing Local Alignments 108
6.5 Overlap Alignments 111
6.6 Alignment of cDNA Sequences with Genomes and Affine Gap Penalties 114
6.7 Gotoh's Algorithm for Affine Gap Penalties 116
6.8 Hirschberg's Space Reduction Technique 120

7. Seeded Alignments 125

7.1 Sensitivity and Specificity 126
7.2 Computing Sensitivity and Specificity 129
7.3 Multiple Hits of Seeds 131
7.4 Gapped Seeds 134
7.5 Chaining and Comparative Genomics 135

	7.6	Design of Highly Specific Oligomers	141
	7.7	Seeds for Computing Mismatch Tolerance	145
		7.7.1 Naive Algorithm	145
		7.7.2 BYP Method	146
	7.8	Partially Matching Seeds	147
	7.9	Overlapping, Partially Matching Seeds	151
8.	Whole Genome Shotgun Sequencing		155
	8.1	Sanger Method	156
		8.1.1 Improvements to the Sequencing Method	159
	8.2	Cloning Genomic DNA Fragments	160
	8.3	Basecalling	163
	8.4	Overview of Shotgun Sequencing	165
	8.5	Lander-Waterman Statistics	170
	8.6	Double-Stranded Assembly	172
	8.7	Overlap-Layout-Consensus	175
		8.7.1 Overlap	175
		8.7.2 Layout	177
		8.7.3 Consensus	180
	8.8	Practical Whole Genome Shotgun Assembly	182
		8.8.1 Vector Masking	183
		8.8.2 Quality Trimming	186
		8.8.3 Contamination Removal	187
		8.8.4 Overlap and Layout	188
		8.8.4.1 Seed and Extend	188
		8.8.4.2 Seeding	190
		8.8.4.3 Greedy Merging Approach	191
		8.8.4.4 Longer Repeat Sequence	192
		8.8.4.5 Iterative Improvements	193
		8.8.4.6 Accelerating Overlap Detection	194
		8.8.4.7 Repeat Sequence Detection	196
		8.8.4.8 Error Correction	199
		8.8.4.9 Repeat Separation	199
		8.8.4.10 Maximal Read Heuristics	201
		8.8.4.11 Paired Pair Heuristics	202
		8.8.4.12 Parallelization	203
		8.8.4.13 Eliminating Chimeric Reads	204
		8.8.5 Scaffolding	205

	8.8.5.1 How Many Mate Pairs Are Needed for	
	Scaffolding?	207
	8.8.5.2 Iterative Improvements	208
	8.8.6 Consensus	210
8.9	Quality Assessment	211
8.10	Past and Future	213

Software Availability   221

Bibliography   223

Index   233

Chapter 1

Simple String Search

Searching a very long sequence for a fairly short query string is a fundamental operation in processing large genome sequences. The query string may occur exactly as it is in the target sequence, or there could be variants that do not match the query exactly, but which are very similar. The former case is called exact matching, while the latter is approximate matching. This chapter describes some simple, naive algorithms for exact matching.

1.1 Storing a String in an Array

Let us consider a search of the target string ATAATACGATAATAA using the query ATAA. It is obvious that ATAA occurs three times in the target, but queries do not always appear in the target; e.g., consider ACGC. Therefore, we need an algorithm that either enumerates all occurrences of the query string together with their positions in the target, or reports the absence of the query.

For this purpose, the target and query strings are stored in a data structure called an *array*. An array is a series of elements, and each element is an object, such as a character or an integer. In order to create an array in main memory, one must declare the number of elements to obtain the space. For example, storing the target string ATAATACGATAATAA requires an array of 15 characters.

To access an element in an array, one specifies the position of the element, which is called the *index*. There are two major ways of indexing. *One-origin* indexing is commonly used and requires that the head element of an array has index one, the second one has index two, and so on. In programming, however, it is more typical to use *zero-origin* indexing, in which the head element has index zero.

zero-origin	0	1	2	3	4	5	6	7	8	9	10	11	12	13	14
one-origin	1	2	3	4	5	6	7	8	9	10	11	12	13	14	15
target	A	T	A	A	T	A	C	G	A	T	A	A	T	A	A

Fig. 1.1 Zero-origin and one-origin indexing.

Figure 1.1 illustrates the difference between the two types of indexing. According to zero-origin (and one-origin indexing, respectively), C is in the 6th (or 7th) position, and the G in the 7th (or 8th) position is the 1st (or 2nd) element from C. In general, the j-th element from the i-th element is the $(i+j)$-th $((i+j-1)$-th) element in terms of zero-origin (one-origin) indexing. If an array has n elements, the last element has index $n-1$ with zero-origin indexing. Although readers who are not familiar with zero-origin indexing may find it puzzling initially, zero-origin indexing is used in this book because most software programs adopt this form.

Let target denote the array in Figure 1.1. The i-th element in the target is specified by target[i]. For example, target[6] and target[7] are C and G in terms of zero-origin indexing.

1.2 Brute-Force String Search

Figure 1.2 presents a simple, naive program called a *brute-force search*, which searches the target string in the target array for all occurrences of the query string in the query array. The Java programming language is used to describe programs in this book [1]. The program checks whether the query string matches the substring of the same length in the target that starts at the i-th position for each i = 0, 1, ... The indexes i and j show the current positions in the target and query that are being compared.

The outer for-loop initializes i to zero, and it increments i until i reaches targetLen - queryLen. Inside the for-loop, j is initially set to zero. The while-loop checks whether the query matches the substring starting from the i-th position in the target string. Each step compares the j-th character in the query with the i+j-th one in the target, and then moves j to the next position if the two letters are equal. This step is iterated until a

---
[1] We adopted the Java programming language because of the availability of Eclipse, an extensible development platform and application framework for building software. We have found Eclipse to be effective teaching material in a Bioinformatics programming course that we have been teaching since 2003. For better computational efficiency, it is straightforward to transform all the Java programs in this book into C/C++ programs.

```java
public static void bruteSearch( int[] target, int[] query ) {
    // "target.length" returns the size of the array target.
    int targetLen = target.length;
    int queryLen = query.length;
    for(int i = 0; i + queryLen <= targetLen; i++){ // i++ means i=i+1.
        int j = 0;
        while(j < queryLen && target[i+j] == query[j])  j = j+1;
        if(j == queryLen)  System.out.println(i);
    }
}
```

query	0	1	2	3											
	A	T	A	A											

target		0	1	2	3	4	5	6	7	8	9	10	11	12	13	14
i+j	j	A	T	A	A	T	A	C	G	A	T	A	A	T	A	A
0,...,3	0,...,3	A	T	A	A											
1	0		A													
2,3	0,1			A	T											
3,...,6	0,...,3				A	T	A	A								
4	0					A										
5,6	0,1						A	T								
6	0							A								
7	0								A							
8,...,11	0,...,3									A	T	A	A			
9	0										A					
10,11	0,1											A	T			
11,...,14	0,...,3												A	T	A	A

Fig. 1.2 The upper half shows a brute-force string search algorithm for scanning the target string for the query. The lower half presents execution of the brute-force string search algorithm on the query and target arrays.

pair of characters disagrees, or all the pairs are equal. In the latter case, j is incremented until it reaches queryLen. Subsequently, the program exits the while-loop and prints the position i at which the query occurs in the target if j equals queryLen. "System.out.println(i)" prints the value of i to the standard output.

To illustrate how the brute-force search program works, consider the target string ATAATACGATAATAA and the query ATAA stored in the arrays target and query in Figure 1.2. The lower half of Figure 1.2 illustrates the execution of the program. The leftmost two columns present the ranges that i + j and j take until the program exits the while-loop. Letters in gray boxes indicate positions at which the target and query disagree. For example, consider the case when i ranges from 3 to 6 and j takes values from 0 to 3. The 6th character, C, in the target and the 3rd character, A, in the query differ, as indicated by the gray box.

1.3 Encoding Strings into Integers

The brute-force string search iterates the step of making pairwise comparisons between individual letters at the same positions in the query string and each substring of the same length in the target sequence. Since the maximum number of character comparisons in each step is determined by the length of the query, handling a fairly long query may increase the overall computation time. Here, we present an idea for improving this basic string comparison step by transforming strings into integer representations, thereby replacing the string comparison with one operation that compares two integers.

We attempt to encode a string into an integer so that the integer can be decoded to the original string, thereby making it possible to compare integer representations instead of strings themselves. Since this book mainly considers strings that are comprised of four nucleotide letters A, C, G, and T, string $b_0 b_1 \ldots b_{k-1} (b_i \in \{A, C, G, T\})$ of length k is called a *k-mer* or a *k-mer string* in what follows. We inductively define a function that maps a k-mer to an integer called a *k-mer integer*. The basic step is to define *encode* for 1-mer that consists of one nucleotide letter. Since four integers are sufficient to provide unique representations for the four individual nucleotide letters, let us associate nucleotide letters with integers in alphabetical order:

$$encode(A) = 0, encode(C) = 1, encode(G) = 2, encode(T) = 3$$

In the inductive step, we extend the definition of *encode* to k-mer $s = b_0 b_1 \ldots b_{k-1} (b_i \in \{A, C, G, T\})$ according to the following formula:

$$encode(s) = \sum_{i=0}^{k-1} 4^{k-1-i} encode(b_i).$$

$encode(s)$ is the k-mer integer. For example, we have

$$encode(\text{ATAA}) = 0 \cdot 4^3 + 3 \cdot 4^2 + 0 \cdot 4^1 + 0 \cdot 4^0 = 48.$$

A k-mer is transformed into $2k$ bits, and its integer values range from 0 to $2^{2k} - 1$. If integers are coded in unsigned 32 bits, this transformation allows us to represent strings of at most 16 nucleotide letters.

Decoding a k-mer integer into the original string obviously involves iterating the step that divides a given integer (or its running quotient) by 4 and prints the letter corresponding to the reminder. The step is iterated

Simple String Search

target	0	1	2	3	4	5	6	7	8	9	10	11	12	13	14
	A	T	A	A	T	A	C	G	A	T	A	A	T	A	A

Encoding substrings into integer representations from left to right

(48) 195 12 49 198 24 99 140 (48) 195 12 (48)

query
ATAA = 48

Checking the equality between each integer and the query integer

```
public static void intBruteSearch( int[] target, int[] query ) {
    // Exit if the target is shorter than the query.
    if(target.length < query.length) return;
    // Generate k-mer integer representations of the target and query.
    int[] intTarget = generateIntTarget(target, query.length);
    int intQuery = generateIntQuery(query);
    // Search intTarget for the query.
    for(int i = 0; i < intTarget.length; i++)
        if(intTarget[i] == intQuery) System.out.print(i+" ");
    System.out.println();
}
public static int[] generateIntTarget( int[] target, int queryLen ){
    // Exit if the target is shorter than the query.
    if(target.length < queryLen) return null;
    int intTargetLen = target.length - queryLen + 1;
    int[] intTarget = new int[intTargetLen];
    // Generate intTarget array.
    int tmp = 0;   int encode = 1;
    for(int i = 0; i < intTargetLen; i++){
        if(i == 0)
            for(int j=0; j<queryLen; j++){ // Initialize variables.
                tmp = 4*tmp + target[j];  encode = encode*4;  }
        else // Compute the next value of tmp from the previous value.
            tmp = 4*tmp- encode*target[i-1] + target[i+queryLen-1];
        intTarget[i] = tmp;
    }
    return intTarget;
}
public static int generateIntQuery(int[] query){
    int intQuery = 0;
    for(int i = 0; i < query.length; i++) intQuery = 4*intQuery + query[i];
    return intQuery;
}
```

Fig. 1.3 The upper half illustrates how the program in the lower half operates on the target string when the query string is given.

k-times for a k-mer integer. For example, if 6 is a 3-mer integer, we perform the following steps to decode 6 into ACG:

Step	Quotient	Reminder	Print
1	$6/4 = 1$	$6 \bmod 4 = 2$	G
2	$1/4 = 0$	$1 \bmod 4 = 1$	C
3	$0/4 = 0$	$0 \bmod 4 = 0$	A

One k-mer integer can be decoded into different strings if k is changed. For example, we can decode the k-mer integer 1 into C, AC, and AAC respectively for $k = 1, 2, 3$. To avoid ambiguity, if the value of k is determined, it should be stated explicitly. However, in the general context in which the value of k should remain open, we will use k-mer integers to express this notion.

The upper half of Figure 1.3 illustrates the procedure used to encode all 4-mer substrings in ATAATACGATAATAA into 4-mer integers. Individual 4-mer integers are generated sequentially and are checked to see if they equal the 4-mer integer of the query ATAA, which is 48. Consider the computation of 195, the 4-mer integer of TAAT. Observe

$$encode(\text{ATAA}) = 0 \cdot 4^3 + 3 \cdot 4^2 + 0 \cdot 4^1 + 0 \cdot 4^0 = 48$$
$$encode(\text{TAAT}) = 3 \cdot 4^3 + 0 \cdot 4^2 + 0 \cdot 4^1 + 3 \cdot 4^0 = 195$$

It is evident that the 4-mer integer of TAAT can be calculated from the value of the previous substring ATAA, i.e.,

$$encode(\text{TAAT}) = (encode(\text{ATAA}) - 0 \cdot 4^3) \cdot 4 + 3 \cdot 4^0.$$

Note that the above formula uses only three arithmetic operations: subtraction, multiplication, and addition, if 4^3 has been computed once previously. In general, for string $s_k = b_k b_{k+1} \ldots b_{k+l-1}$ and $s_{k+1} = b_{k+1} b_{k+2} \ldots b_{k+l}$, we have

$$encode(s_{k+1}) = (encode(s_k) - encode(b_k) \cdot 4^{l-1}) \cdot 4 + encode(b_{k+l}).$$

Therefore, even if longer substrings are processed, it holds that three arithmetic operations are required before moving on to handle the next substring, which is more efficient than scanning both the query string and each substring in the target. The lower half of Figure 1.3 presents a program that implements the brute-force string search algorithm empowered with k-mer integers of substrings.

1.4 Sorting k-mer Integers and a Binary Search

To search the target string for one query string, scanning `target` from the head to the tail for each query is not a demanding task. However, if a large number of queries has to be processed and the target string is extremely long, like the human genome sequence, we need to accelerate the overall performance. Here, we introduce a technique for preprocessing the target sequence, so that the task of searching for one query string can be done more efficiently at the expense of using more main memory.

The upper half of Figure 1.4 shows a table in which the 4-mer integer for the substring starting at the i-th position is put into `intTarget[i]`. The size of `intTarget` is `targetLen - queryLen + 1`, which is denoted by `intTargetLen` in what follows. Occurrences of `ATAA` in the target string can be found by searching `intTarget` for the 4-mer integer of `ATAA`, 48. However, putting all the 4-mer integers into an array does not improve the performance.

Our key idea is to sort the values of `intTarget` in ascending order, as illustrated in the lower half of Figure 1.4. Basic efficient algorithms for sorting a list of values will be introduced in Chapter 2. Simple application of the sorting process may lose information regarding the location of each element `intTarget`. In order to memorize the position (index) of each element, we use another array named `posTarget`. For example, suppose that after `intTarget` has been sorted, `intTarget[i]` is assigned to the j-th position of `intTarget*`, where `intTarget*` denotes the result of sorted `intTarget`. Then, we simultaneously assign i to `posTarget[j]`.

In a sorted list of values, a query value can be located quickly using a binary search. A binary search probes the value in the middle of the list, compares the probed value and the query to decide whether the query is in the lower or upper half of the list, and moves to the appropriate half to search for the query. For example, let us search the sorted `intTarget*` for 24. The binary search calculates a middle position by dividing 0+11 by 2 and obtains the quotient 5. The search then probes the 5-th position and finds that its value, 48, is greater than 24. Therefore, it moves to the lower half and probes the 2nd position $((0 + 5)/2 = 2)$ to identify 24. In order to memorize the 5th position, at which 24 is located in the original `intTarget` array, `posTarget[2]` stores the position 5.

When the binary search processes a list of n elements, each probing step narrows the search space to its lower or upper half. After k probing steps, the binary search focuses on approximately $n/2^k$ elements, and it stops

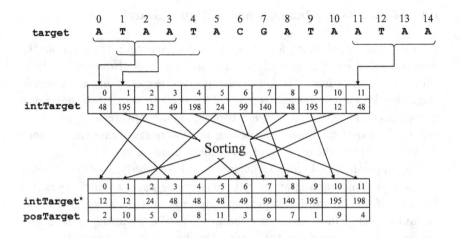

Fig. 1.4 The 4-mer substrings in the top array target are transformed into 4-mer integers in the middle array intTarget. The elements in intTarget are sorted in ascending order and are put into intTarget*. posTarget memorizes the original positions of the elements in intTarget before they are sorted.

probing when two elements remain to be investigated. Solving $n/2^k = 2$ shows that k is almost equal to $\log_2 n - 1$, which indicates the approximate number of probing steps that the binary search needs to take. For instance, when $n = 2^{30} \approx 10^9$, $k \approx 29$.

1.5 Binary Search for the Boundaries of Blocks

The binary search described above works properly when the sorted list contains no duplicates. However, the sorted intTarget* in Figure 1.4 contains multiple occurrences of the same numbers, and the program is required to output all the positions of occurrences of the query. For example, 48 appears from positions 3 to 5 successively in the sorted intTarget*, and it occurs at positions 0, 8, and 11 in the original array, as posTarget indicates. Since occurrences of the query 4-mer integer are successive in the sorted list, it is sufficient to calculate the leftmost and rightmost positions of the successive occurrences, and each of the two boundary positions can be calculated in the binary search program given in Figure 1.5.

In the binary search for the leftmost position, the program uses the variable left to approach the leftmost index by scanning the array from the left, as illustrated in Figure 1.5. In order to implement this idea,

```
public static void binarySearch( int[] target,  int[] query ){
    // Exit if the target is shorter than the query.
    if(target.length < query.length) return;
    // Generate k-mer integer representations of target and query.
    int[] intTarget = generateIntTarget(target, query.length);
    int intQuery = generateIntQuery(query);
    // Sort elements in intTarget and
    // memorize the original positions in posTarget.
    int targetLen = intTarget.length;
    int[] posTarget = new int[targetLen];
    for(int i=0; i<targetLen; i++) posTarget[i]=i;
    randomQuickSort(intTarget, posTarget, 0, targetLen-1);
    // Search for the left boundary.
    int left, right, middle;
    for(left=0, right=targetLen; left < right; ){
        middle = (left + right) / 2;
        if( intTarget[middle] < intQuery) left = middle + 1;
        else right = middle; }
    int leftBoundary = left;
    // Search for the right boundary.
    // We use "right" to approach the index next to the right boundary.
    for(left=0, right=targetLen; left < right; ){
        middle = (left + right) / 2;
        if( intTarget[middle] <= intQuery ) left = middle + 1;
        else right = middle; }
    // Print positions in the range between the two boundaries.
    for(int i = leftBoundary; i < right; i++)
        System.out.print(posTarget[i]+" ");
    System.out.println();
}
```

	0	1	2	3	4	5	6	7	8	9	10	11	12
intTarget*	12	12	24	48	48	48	49	99	140	195	195	198	
leftmost	l_0		l_1	l_2									
				r_2			r_1						r_0
rightmost	l_0			l_1			l_2						
							r_1						r_0

Fig. 1.5 The upper half presents a binary search algorithm for seeking the leftmost and rightmost boundary positions of contiguous occurrences of the query k-mer integer. The lower half shows the operation of the program on the sorted intTarget* for the query 4-mer integer, 48.

the program sets **left** to the position next to the middle position when the query 4-mer integer is higher than the target 4-mer interger in the middle. In the leftmost and rightmost search rows, l_0 and r_0 show the initial positions of **left** and **right**, respectively, while l_i and r_i indicate the i-th updated values during the execution. When the program exits the for-loop, observe that `left` contains the leftmost index where the query 4-mer integer appears in the leftmost search of Figure 1.5. In contrast, when searching for the rightmost location, `right` is intended to store the position next to the rightmost position in the final step.

Chapter 2

Sorting

We present several sorting algorithms that take an array of integers and output an array in which all the elements are sorted in ascending order. Sorting algorithms may appear to be outside the scope of processing large-scale genome sequences, but they are indispensable for designing fast string search algorithms. In addition, sorting algorithms are useful for understanding several fundamental ideas concerning algorithm design.

2.1 Insertion Sort

The insertion sort is one of the simplest sorting algorithms. It scans individual elements in the target array from the left and exchanges the focal element with the element to its left if the former is less than the latter. The upper picture in Figure 2.1 illustrates how the insertion sort operates. First, it confirms that the second element (195) is greater than the first element (48). Then, it focuses on the third element (12), recognizes that 12 is less than the second element (195), and exchanges these two elements. Subsequently, it switches 12 and 48. The solid lines show the movement of focal elements to their proper positions, and the dotted lines represent how the other elements are shifted one to the right when the focal elements are moved to the left.

The Java program in the lower part of Figure 2.1 implements the insertion sort. For each $i = 2, 3, \ldots$, the program sorts the first i elements by iterating the process of scanning the first $i - 1$ elements, which were sorted in the previous step, from the right in order to locate the i-th element in the proper position so that first i elements are sorted. The insertion sort stops shifting the focal element if the element is not less than the element to its left. This rule is valid because all the elements to the left of the focal

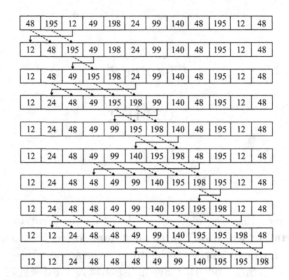

```
public static void insertionSort(int[] target){
    int n = target.length;
    for(int i = 1; i < n; i++){
        int v = target[i]; int j;
        for(j = i; 0 < j && v < target[j-1]; j--) target[j] = target[j-1];
        target[j] = v;
    }
}
```

Fig. 2.1 The upper picture illustrates how the insertion sort works, and a Java program that codes the insertion sort is shown below.

element have already been scanned and sorted. The number of movements varies. For instance, in the upper picture, the second element (195) is not moved, but the second to last element (12) is shifted to the second position. The number of exchanges is maximized when the input is in descending order because each focal element must be shifted to the leftmost position. Therefore, in this worst case, the number of exchanges is $(n-1)n/2$ when the input length is n. Nevertheless, insertion sort is the method of choice when the input is very short, for example, when the length is less than ten.

2.2 Merge Sort

The merge sort utilizes a *divide-and-conquer* approach to partition the original problem into subproblems, solve individual subproblems, and integrate

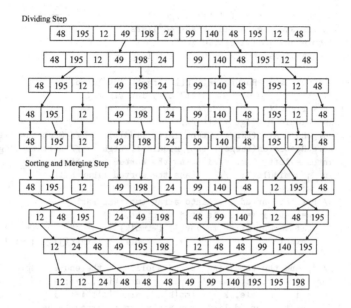

Fig. 2.2 Illustration of how the merge sort works.

the sub-solutions to answer the input problem. The merge sort is one of the oldest sorting algorithms. In 1945, John von Neumann prepared a merge sort program to test some code for the EDVAC computer because the developers of EDVAC were particularly interested in using the machine for nonnumerical applications. For more on the historical development of sorting algorithms, see Knuth's book on sorting [48].

The merge sort divides the target array into two blocks of nearly half size and iterates this process recursively until each block contains one element. After this dividing step, the algorithm merges two blocks of one element into one sorted block of two elements. It continues the step of merging two sorted blocks into one sorted block recursively until it outputs the sorted block for the target array. Figure 2.2 illustrates how the running example array is sorted using the merge sort.

Figure 2.3 presents a Java program that implements the merge sort. It first divides the target array into halves recursively, and then it merges two sorted blocks. Subsequently it uses two variables, i and j, to scan the left and right sorted blocks. Two elements, target[i] and target[j], in the two blocks are compared; the smaller element is added to the end of the temporary array named tmp and the element next to the smaller element is

```
public static void mergeSort( int[] target ){  // target != null.
    // Used for storing partially sorted lists.
    int[] tmp = new int[target.length];
    mergeRecursion(target, tmp, 0, target.length-1);
}
public static void mergeRecursion(int[] target, int[] tmp,
                        int left, int right){
    if(left < right){    // No data movements if left == right.
        int middle = (right+left)/2;
        mergeRecursion(target, tmp, left, middle);      // Sort the left half.
        mergeRecursion(target, tmp, middle+1, right);   // Sort the right half.
        int i = left;         // i scans the sorted left half.
        int j = middle+1;     // j scans the sorted right half.
        int k;
        // Put the sorted list into array tmp temporarily.
        for(k=left; i <= middle && j <= right; k++){
            if(target[i] <= target[j]){ tmp[k] = target[i];  i++; }
            else{                       tmp[k] = target[j];  j++; }
        }
        // Append the remaining sorted sublist to the end of tmp.
        for(; j <= right;   k++){ tmp[k] = target[j]; j++; }
        for(; i <= middle;  k++){ tmp[k] = target[i]; i++; }
        // Copy the sorted list from array tmp to array target.
        for(k = left; k <= right; k++ ) target[k] = tmp[k];
    }
}
```

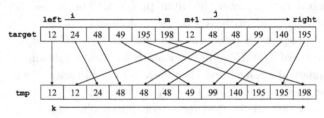

Fig. 2.3 The upper portion presents a Java program for dividing a target array into blocks and merging sorted blocks, and the lower portion shows how the program merges two sorted blocks into one.

considered in the following step. The lower picture in the figure illustrates the process. Note that while merging the two sorted blocks, we can search both blocks from left to right without backtracking because we never hit elements smaller than ones that have been scanned already.

Therefore, given two sorted blocks of length l, it requires at most $2l$ data comparisons and $2l$ data movements from target to tmp to merge the blocks to yield one sorted block. Since the number of data movements is no less than the number of data comparisons, the former factor dominates the computational performance of the merge sort. For a target array of length $n (\geq 1)$, let $C(n)$ denote the number of data movements from target to tmp to sort the input. Then, we have

$$C(n) = \begin{cases} 0 & n = 1 \\ 2C(n/2) + n & n > 1, n \text{ is even.} \\ C(\lfloor n/2 \rfloor) + C(\lfloor n/2 \rfloor + 1) + n & n > 1, n \text{ is odd.} \end{cases}$$

If $n = 1$, target has a single element, it does not make sense to move the element from target to tmp in order to sort target; consequently, the program moves no characters. Therefore, the cost $C(1)$ equals zero. In this formula, $\lfloor n/2 \rfloor$ means the greatest integer that is lower than or equal to $n/2$. Next, observe that

$$\frac{C(2^k)}{2^k} = \frac{C(2^{k-1})}{2^{k-1}} + 1 = \frac{C(2^{k-2})}{2^{k-2}} + 2 = \frac{C(2^0)}{2^0} + k = k,$$

and therefore $C(2^k) = 2^k k$. Moreover, note that $C(n)$ increases monotonically as n increases. For n such that $2^k \leq n < 2^{k+1}$,

$$2^k k = C(2^k) \leq C(n) < C(2^{k+1}) = 2^{k+1}(k+1).$$

From $n < 2^{k+1}$, we have $n/2 < 2^k$ and $\log_2 n - 1 < k$. From $2^k \leq n$, we have $k \leq \log_2 n$. Consequently,

$$(n/2)(\log_2 n - 1) < 2^k k \leq C(n) < 2^{k+1}(k+1) \leq 2n(\log_2 n + 1).$$

Therefore, $n \log_2 n$ roughly approximates $C(n)$.

Although the above running time analysis mainly considers the dominant factor $C(n)$, the number of data movements, the program also executes other less dominant statements, such as comparing characters, setting variables to values, and calling functions recursively. However, it is difficult to estimate the number of data comparisons precisely because the number is largely dependent on the input data. For example, consider what happens if the input is sorted in ascending or descending order. The number of character companions in the former case is much smaller than that in the latter case. Although it would be helpful to clarify the running time of the algorithm in the worst-case scenario, it is not that simple to make an accurate estimate in practice. We can at least state that the running time is bounded by a constant factor of the dominant factor $C(n)$ if the constant is sufficiently large. We will continue to discuss the worst-case running time analysis in a more general setting.

2.3 Worst-Case Time Complexity of Algorithms

To estimate the computational performance of algorithms, it is widely accepted that one considers the worst-case scenario, i.e., when an algorithm performs worst at handling the given data because the worst-case execution time analysis guarantees that the algorithm will never run longer. For example, in the analysis of the merge sort, for n such that $2^k \leq n < 2^{k+1}$,

$$C(n) < 2n(\log_2 n + 1).$$

The upper bound $2n(\log_2 n + 1)$ has a complex form, so that alternative upper bounds in simpler form, e.g., $4n \log_2 n$, may be preferable to provide a rough estimate of the worst-case running time. However, the use of a loose upper bound, e.g., n^2, should be avoided because the difference between $n \log_2 n$ and n^2 is not within a fixed constant factor for an asymptotically large n, such that $n > n_0$ for some n_0. It would be reasonable to select an upper bound with a simple form within a constant factor for the worst-case running time.

Definition 2.1 Let $f(n)$ and $g(n)$ denote functions that input nonnegative integers n and output nonnegative real numbers. Suppose that $g(n)$ is bounded by a function $f(n)$ within a constant factor for asymptotically large n, i.e., $g(n) \leq cf(n)$ ($n_0 \leq n$) for some constants c and n_0. $f(n)$ is called an *asymptotic upper bound* of $g(n)$. Let

$$O(f(n)) = \{g(n) \mid f(n) \text{ is an asymptotic upper bound of } g(n)\}.$$

$O(f(n))$ is called the *O-notation* of $f(n)$. The worst-case time complexity of an algorithm is called $O(f(n))$ if its worst-case running time is in $O(f(n))$.

The O-notation gives us a concise expression for the complexity of the worst-case computational time. For example, $C(n) \leq 2n(\log_2 n + 1) \in O(n \log_2 n)$. Since $C(n)$ dominates the total running time of other operations in the merge sort within a constant factor, we assert that the worst-case time complexity of the merge sort is $O(n \log_2 n)$.

Other O-notations, such as $O(1)$, $O(\log_2 n)$, $O(n)$, and $O(n^2)$, will be used in this book. The time complexity of an algorithm is $O(1)$ if the algorithm can process a request within a constant time, even if the input size n gets large. If the worst-case running time of an algorithm is proportional to the size of the input n, its time complexity is described using $O(n)$. The worst-case time complexity of the insertion sort is $O(n^2)$.

2.4 Heap Sort

The merge sort is an optimal sorting algorithm in terms of the worst-case time complexity. It is used to handle large data in external memory, but to sort data that fit in main memory, faster algorithms are available. One more efficient algorithm is the heap sort. The worst-case time complexity of the heap sort is also $O(n \log_2 n)$, but it is empirically faster than the merge sort because the number of comparisons and data exchanges executed in the heap sort is often much smaller than that in the merge sort.

The heap sort uses a data structure called a *heap*. A heap is viewed as a binary tree structure with nodes that contain the elements in the target array. Any heap needs to satisfy the *heap property*: the value of any node, except the root node, must be lower than or equal to the value of its parent. For example, Figure 2.4 illustrates a heap for the target array, and the individual nodes meet the heap property. Although there could be more than one heap that represents the elements in the target array, the maximum value is always stored at the root. Otherwise, the root would have a value that is less than the maximum value. Consider the path from the root to a node with the maximum value. Given the heap property, the value of the root would then be greater than the maximum, which is a contradiction.

In order to store a heap in a data structure, the values in the heap are put into an array so that the value of the root is assigned to the first position, and the other values are put into the array according to breadth-first ordering of the nodes. More precisely, define the *level* of a node as the

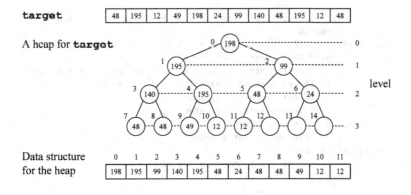

Fig. 2.4 Data structure of a heap.

number of edges on the path from the root to the node. In breadth-first ordering, the root node is ordered first, nodes of level i are ordered before those of deeper level $j\ (> i)$, and nodes of the same level are ordered from left to right in the heap. For example, the numbers associated with the nodes in Figure 2.4 illustrate breadth-first ordering, and the values in the nodes are stored in the bottom array conforming to the ordering.

Next, we describe how to build a heap for the target array. We simply scan elements in the target array from the left, and put an individual element at the end of the heap. However, the appended node may violate the heap property. In this case, we exchange the element in the appended node with that in its parent to guarantee the heap property. However, the revised parent node might still not meet the property, and we then have to iterate the process of moving the appended element upward until the heap property is ensured. Let us call this procedure *upheap*. Figure 2.6

```
public static void heapSort(int[] target){ // target != null.
    int heapSize = target.length;
    int[] heap = new int[heapSize];
    // Place all elements in the target array in the heap.
    for(int i=0; i<heapSize; i++)  upheap(heap, i, target[i]);
    // Remove the elements at the root node one at a time.
    for(int i = 0; i < heapSize; i++){
        target[heapSize-1-i] = heap[0];
            // Remove the largest element from the root.
        heap[0] = heap[heapSize-1-i];
            // Move the last element in the heap to the root.
        downheap(heap, 0, heapSize-1-i-1);
            // Execute downheap to maintain the heap.
    }
}
public static void upheap(int[] heap, int leafPosition, int element){
    int child, parent;
    for( child = leafPosition, parent = (child-1)/2; // Initialization.
         0 < child && element > heap[parent];
            // Confirm that the child has its parent and check whether
            // the element is greater than the value of the parent.
         child = parent, parent = (child-1)/2 )
            // Move the child and parent upward.
        heap[child] = heap[parent];
            // Move the parent's value to the child.
    heap[child] = element;
            // Locate the element in the proper position.
}
```

Fig. 2.5 A Java program that codes the heap sort and the upheap procedure.

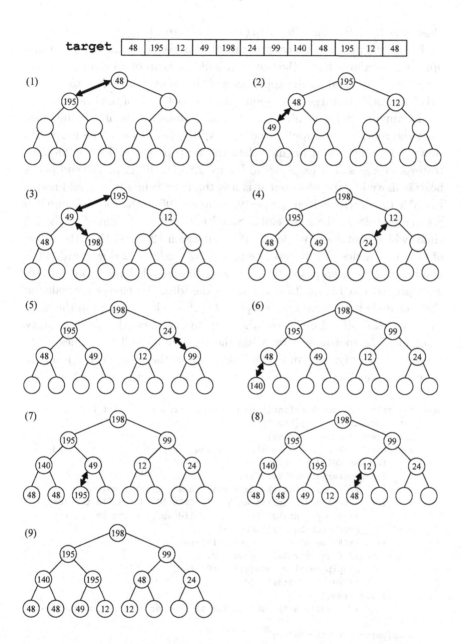

Fig. 2.6 Placing data in a heap. Diagrams (1) to (9) show how to maintain a heap by exchanging elements in nodes that violate the heap property with elements in their parent nodes. Double arrows indicate data exchanges.

illustrates the iteration of inserting elements into the heap.

Figure 2.5 presents a Java program that implements the heap sort and upheap procedure. First, the program builds a heap of all elements in the target array by iterating the application of the upheap procedure to individual elements in the target. Consequently, the root node of the heap contains the maximum. Subsequently, the program eliminates the maximum value from the root, and it supplies the value of the last node in the heap with that of the root, which might violate the heap property. Figure 2.8 illustrates such cases. For example, in Figure 2.8 (1), the 12 at the end of the heap is moved to the root, but it is less than the values in its child nodes. In order to ensure the heap property, we move 195 to the root because it is greater than 99 in the right child. Simultaneously, 12 is moved to the left child, which contained previously 195, but again 12 is less than the values of the child nodes. We iterate the process of pushing 12 downward, while maintaining the heap property. This procedure is called *downheap*, and the Java program in Figure 2.7 implements the idea. Whenever the value at the root node is removed and is replaced with the last element in the heap, the downheap procedure revises the heap to guarantee the heap property. Therefore, the root node always has the maximum of all the values in the heap. In this way, we can extract elements in the target array from the heap in descending order, allowing us to put each element back into the

```
public static void downheap(int[] heap, int startPosition, int heapLast){
    int startValue = heap[startPosition];
    int parent = startPosition;
    while(parent*2+1 <= heapLast){        // The parent has the left child.
        int leftChild = parent*2+1;
        int greaterChild = leftChild;
            // Assume that the left child has a greater value.
        if( leftChild < heapLast && heap[leftChild] < heap[leftChild+1])
            // The parent has the right child whose value is greater.
            greaterChild = leftChild+1;
        if(startValue < heap[greaterChild]){
            // Push startValue downward.
            heap[parent] = heap[greaterChild];
            parent = greaterChild;
        }else break;
            // Terminate the search for the proper position of startValue.
    }
    heap[parent] = startValue;
}
```

Fig. 2.7 Downheap.

original target array from its end to output the sorted list of the target.

In a heap with n elements, the level of any node is at most $\log_2 n$. Therefore, the number of data movements in each upheap/downheap operation is $O(\log_2 n)$. Since $2n$ upheap/downheap operations are made, the worst-case time complexity of the heap sort is $O(n \log_2 n)$.

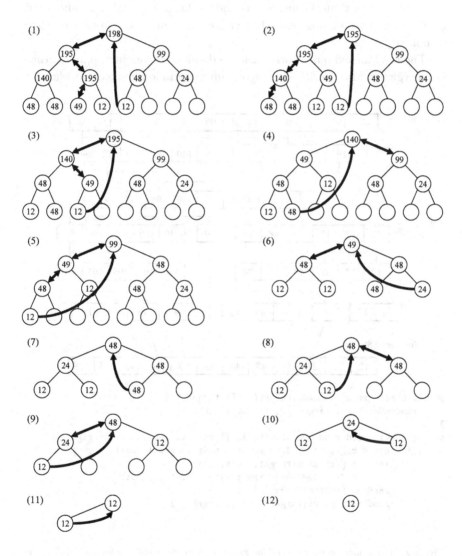

Fig. 2.8 Deleting data from a heap. After the last element in the heap has been moved to the root, the heap property is maintained using the downheap procedure.

2.5 Randomized Quick Sort

Here, we present another sorting algorithm called the *randomized quick sort* that Hoare proposed in 1962 [37]. Although the worst-case time complexity of this algorithm is $O(n^2)$, it is able to handle most real examples empirically much faster than the merge sort and the heap sort, which are supposed to be superior to the randomized quick sort in terms of the worst-case time complexity measurement.

The randomized quick sort takes the divide-and-conquer approach that the merge sort also adopts. The algorithm first randomly selects an element

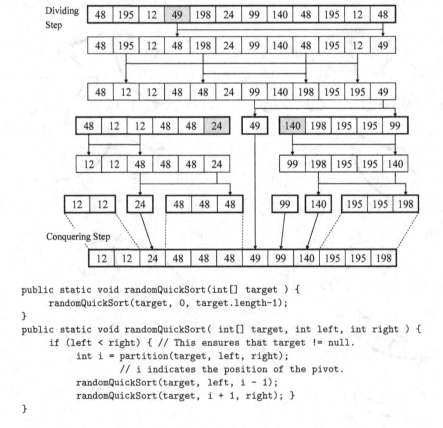

```
public static void randomQuickSort(int[] target ) {
    randomQuickSort(target, 0, target.length-1);
}
public static void randomQuickSort( int[] target, int left, int right ) {
    if (left < right) { // This ensures that target != null.
        int i = partition(target, left, right);
            // i indicates the position of the pivot.
        randomQuickSort(target, left, i - 1);
        randomQuickSort(target, i + 1, right); }
}
```

Fig. 2.9 The upper picture shows the general concept of the randomized quick sort algorithm. Gray boxes indicate pivots that are selected randomly. Arrays outlined in bold lines indicate the input, the output of the dividing step, and the final sorted array.

in the target array called the *pivot*, and then divides the array into three blocks so that the middle block contains the pivot only; values lower than the pivot are moved to the left block and values higher than the pivot are put in the right block. Since this step has not sorted the left or right blocks yet, the algorithm applies the step to both of these blocks recursively until it partitions the original array into singleton blocks that contain only one element. As the pivots are sorted in ascending order, combining singleton blocks gives the sorted list of the target array. Figure 2.9 illustrates the concept of the randomized quick sort algorithm.

The lower part of Figure 2.9 presents the part of the randomized quick sort algorithm that partitions the input sequence. Function `partition` randomly selects the pivot and moves values lower than the pivot to the left block, while moving values greater than the pivot to the right. Then, the algorithm applies this process of moving elements recursively to smaller blocks. The crux of the algorithm is how to implement `partition`. Figure 2.10 presents the Java code of `partition`, and the lower part in the figure illustrates execution of the code.

The program selects the index of a pivot between `left` and `right` at random and sets the pivot to variable `pivot`. It exchanges the pivot and last element of the array. Variables i and j are used to scan the array from the left and right, respectively, to seek `target[i]` ($\geq$ `pivot`) from the left and `target[j]` ($\leq$ `pivot`) from the right. After such elements are identified, they are exchanged. The algorithm iterates this process until i exceeds j, indicating that no more exchanges are necessary. Since `target[i]` $\geq$ `pivot`, we can safely exchange these two values so that the right block has values that are greater than or equal to `pivot`.

Figure 2.9 presents an ideal case when randomly selected pivots are close to the medians of the given arrays, making it possible to produce left and right blocks of almost equal size. The concept of the randomized quick sort works even if the minimum or maximum value is selected as the pivot, as shown in Figure 2.11. Although the left or right block may contain some occurrences of the pivot, recursive application of the concept eventually sorts the target array.

Let us analyze the computational complexity of the randomized sort algorithm. Function `partition` compares `pivot` with each element in the block ranging from `left` to `right`. The worst-case scenario is that given an array in which all elements are distinct, the randomized quick sort algorithm always selects the maximum value as the pivot so that the left block has all elements except the pivot, and the right block is empty. In this case, if

```
public static int partition( int[] target, int left, int right ) {
    // Math.random() selects a random real number x such that 0 <= x < 1.
    // Math.floor(x) returns the greatest integer y such that y <= x.
    int random = left + (int)Math.floor(Math.random()*(right-left+1));
    // Exchange target[random] and target[right].
    int tmp = target[right];   target[right] = target[random];
    target[random] = tmp;
    int pivot = target[right];
    int i = left-1;   // i scans the array from the left.
    int j = right;    // j scans the array from the right.
    for (;;) {
        // Move from the left until hitting a value no less than the pivot.
        for(i++; target[i] < pivot; i++){}
        // Move from the right until hitting a value no greater than the pivot.
        for(j--; pivot < target[j] && i < j; j--){}
        if (i >= j)  break;
        // Exchange target[i] and target[j].
        tmp = target[i];  target[i] = target[j];  target[j] = tmp;
    }
    // Exchange target[i] and target[right].
    tmp = target[i];  target[i] = target[right];  target[right] = tmp;
    return i;
}
```

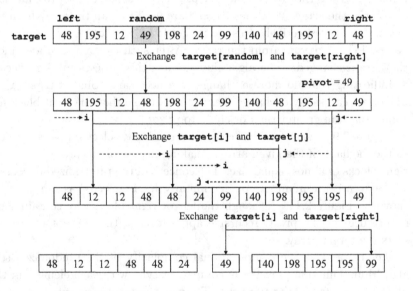

Fig. 2.10 The randomized quick sort program is shown on top, and the lower figure shows how the program divides the target array into three blocks.

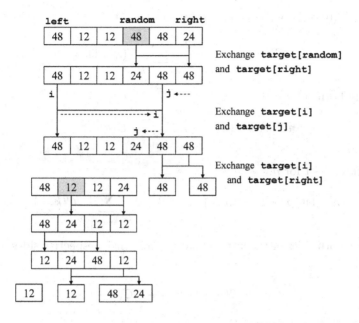

Fig. 2.11 The figure presents a difficult example in which the randomized quick sort algorithm selects the maximum or minimum value as the pivot in each dividing step.

the target contains n elements, the algorithm generates left blocks of size $n-1, n-2, \ldots, 1$ sequentially. Since all the elements in the left blocks are compared with each pivot, the total number of data comparisons is $(n-1) + (n-2) + \ldots + 1 = n(n-1)/2$. However, the probability that the worst case happens is $1/n!$.

Therefore, we are interested in the expected number of data comparisons, and we will show that the number is almost equal to $2n\log_e n$. Suppose that $C(n)$ denote the expected number of comparisons made using function partition when the input is of size n. It is obvious that $C(0) = C(1) = 0$. Suppose that n elements are divided into a left block of size $k - 1 (k \leq 1)$, the singleton middle block for the pivot, and a right block of size $n - k$. The probability of generating such a division is $1/n$. The expected numbers of data comparisons involved in sorting blocks of size $k-1$ and $n-k$ are $C(k-1)$ and $C(n-k)$, respectively. We need $n-1$ comparisons to create such a division. Therefore, $C(n)$ meets:

$$C(n) = n - 1 + \frac{1}{n} \sum_{k=1,\ldots,n} (C(k-1) + C(n-k))$$

Because $C(0) + \ldots + C(n-1) = C(n-1) + \ldots + C(0)$,

$$C(n) = n - 1 + \frac{2}{n} \sum_{k=1,\ldots,n} C(k-1)$$

Multiply both sides by n.

$$nC(n) = n(n-1) + 2 \sum_{k=1,\ldots,n} C(k-1)$$

Substitute $n-1$ for n.

$$(n-1)C(n-1) = (n-1)(n-2) + 2 \sum_{k=1,\ldots,n-1} C(k-1)$$

Subtract both sides of the latter equation from their respective sides of the former.

$$nC(n) = 2(n-1) + (n+1)C(n-1)$$

Divide both sides by $n(n+1)$.

$$\frac{C(n)}{n+1} = \frac{2(n-1)}{n(n+1)} + \frac{C(n-1)}{n} \leq \frac{2}{n} + \frac{C(n-1)}{n} \leq$$

$$\leq \frac{2}{n} + \frac{2}{n-1} + \frac{C(n-2)}{n-1} \leq \ldots \leq \sum_{k=1,\ldots,n} \frac{2}{k} + C(0) \approx 2 \int_1^n \frac{1}{x} \approx 2 \log_e n$$

Consequently, we see that $2n \log_e n$ approximates $C(n)$, indicating $C(n) \in O(n \log_e n)$. This implies that the randomized quick sort algorithm works efficiently on average, although it may need $n(n-1)/2$ data comparisons in the worst case, and hence the worst-case time complexity is $O(n^2)$. In reality, the randomized quick sort is typically faster than the merge sort partly because the randomized quick sort moves elements less frequently than the merge sort.

In Section 1.4, we used a randomized quick sort program that memorizes the original positions of individual elements in an array `posTarget`. A slight modification of the program in Figure 2.10 yields the program shown in Figure 2.12. Variable `int[] posTarget` is added to the argument list and, whenever values in `target` are exchanged, the corresponding two values in `posTarget` are also swapped.

```
public static void randomQuickSort(int[] target, int[] posTarget,
                                   int left, int right ) {
    if (left < right) {
        int i = partition(target, posTarget, left, right);
        randomQuickSort(target, posTarget, left, i - 1);
        randomQuickSort(target, posTarget, i + 1, right);
    }
}
public static int partition( int[] target, int[] posTarget,
                             int left, int right ) {
    int random = left + (int)Math.floor(Math.random()*(right-left+1));
    int tmp = target[right];  target[right] = target[random];
    target[random] = tmp;
    tmp = posTarget[right];   posTarget[right] = posTarget[random];
    posTarget[random] = tmp;
    ...(omitted)...
    for (;;) {
        ...(omitted)...
        tmp = target[i];  target[i] = target[j];  target[j] = tmp;
        tmp = posTarget[i];  posTarget[i] = posTarget[j];
        posTarget[j] = tmp;
    }
    tmp = target[i];  target[i] = target[right];  target[right] = tmp;
    tmp = posTarget[i];  posTarget[i] = posTarget[right];
    posTarget[right] = tmp;
    return i;
}
```

Fig. 2.12 A randomized quick sort for memorizing the original positions. Added statements and their previous statements are displayed.

2.6 Improving the Performance of Quick Sort

Sorting plays important roles in a variety of software programs. Since the randomized quick sort empirically outperforms other sorting algorithms such as the heap sort and merge sort, further improvements on the randomized quick sort have been proposed.

First, let us consider how to select better pivots efficiently. Ideally, one selects the median of all elements for the purpose of dividing the input into halves, but this requires much computation. One typical acceleration method simply uses the median of the first, middle, and last elements in the input as a pivot. The quick sort algorithm that selects pivots in this way is called the *median-of-3* quick sort. The median-of-3 quick sort produces good divisions in many cases; however, its worst-case time complexity is $O(n^2)$. Consider sorting the following example using a median-of-3 quick sort that selects the $n/2$-th position, where $n/2$ is the integer quotient, as the middle position:

| 1 | 5 | 3 | 7 | 2 | 4 | 6 | 8 |

In the initial step, the 4-th position in terms of zero-origin indexing is selected as the middle; hence, the median of 1, 2, and 8 is 2. This produces blocks of size one and six. One can generalize this example to design a pathological one of length n divisible by 4 such that the median-of-3 quick sort is forced to process subarrays of length $n-2, n-4, \ldots, 2$ successively, making the time complexity of the sort $O(n^2)$. This problem is left as an exercise. To reduce the possibility of such a pathological case, we can select three different elements at random and use the median as the pivot instead of using a single element at random. This procedure cuts the expected number of comparisons from $2n \log_e n$ to $(12/7)n \log_e n$ [48].

```
public static void randomQuickSort(int[] target, int aLeft, int right) {
    int left = aLeft;
    while (left < right) {
        int i = partition(target, left, right);
        randomQuickSort(target, left, i-1);
        left = i+1;
    }
}
```

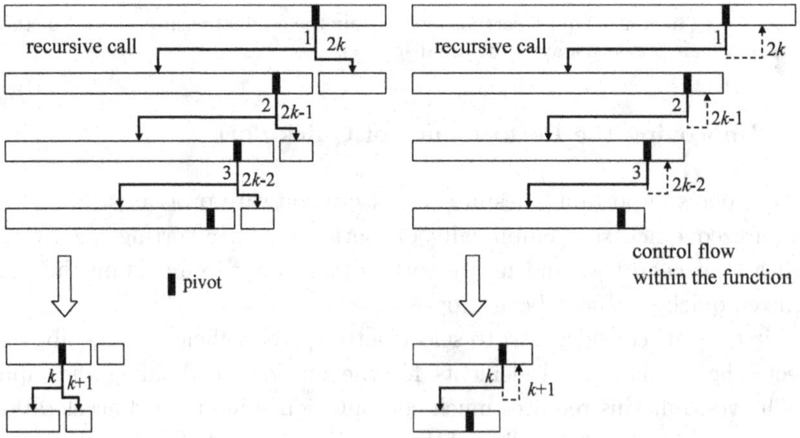

Fig. 2.13 The upper figure shows the randomized quick sort after the unnecessary tail recursion has been eliminated from the original sort in Figure 2.9. Beneath it, the left- and right-hand figures illustrate the behaviors of the original and revised programs. Numbers indicate the execution ordering.

The next improvement is to eliminate unnecessary recursive calls. The randomized quick sort program in Figure 2.9 makes two recursive calls to itself, but the second recursive call, which is called *tail recursion*, is unnecessary because the tail recursion can be rewritten by iteration, as shown in Figure 2.13. This technique is called *tail recursion elimination*, which is done automatically by elaborate compilers. In Figure 2.13, observe that one recursive call to itself is made in the revised program, and the while-loop iteration is used to call itself recursively. The lower two pictures illustrate the difference between the original and revised quick sort algorithms.

Recursive calls are memorized in a waiting list called the *stack* according to the order that they are invoked. These requests are processed in a last-in-first-out manner. The most recent recursive call is pushed on the stack, and it is popped from the stack after it is executed. The number of requests in a stack is called the *stack depth*. The stack depth might be very large, as illustrated in the operations in Figure 2.13, whereas the stack space is usually limited. If the stack depth exceeds the available stack size, the system aborts the execution of the program. Therefore, care has to be taken to reduce the stack depth during execution. One typical way is to handle the shorter interval, divided by the selected pivot, using a recursive

```
public static void randomQuickSort(int[] target, int aLeft, int aRight) {
    int left = aLeft; int right = aRight;
    while (left < right) {
        int i = partition(target, left, right);
        if( i - left <= right - i ){ // The left interval is shorter.
            randomQuickSort(target, left, i-1); left=i+1;
        }else{                       // The right interval is shorter.
            randomQuickSort(target, i+1, right); right=i-1; }
    }
}
```

Fig. 2.14 The upper program makes recursive calls for shorter intervals in order to reduce the stack depth significantly. The lower figure illustrates how the program works.

```
public static void randomQuickSort(int[] target,
                                   int aLeft, int aRight, int minSize) {
    int left = aLeft; int right = aRight;
    while (left + minSize < right) {
        int i = partition(target, left, right);
        if( i - left <= right - i ){
            randomQuickSort(target, left, i-1, minSize);   left = i+1;
        }else{
            randomQuickSort(target, i+1, right, minSize)   right = i-1; }
    }
    // Insertion sort the target array of the range [left, right]
    for(int i = left+1; i <= right; i++){
        int v = target[i];   int j;
        for(j = i; 0 < j && v < target[j-1]; j--) target[j] = target[j-1];
        target[j] = v;
    }
}
```

Fig. 2.15 The randomized quick sort algorithm that incorporates insertion sort for sorting short intervals of length at most minSize, tail recursion elimination, and stack depth reduction.

call. Since the size of the shorter interval is less than half of the original size, the stack depth can be at most $\log_2 n$ when the input length is n. The upper part of Figure 2.14 shows a Java program that implements this idea, and the lower part illustrates how the program works.

Another standard optimization technique is to switch to insertion sort when handling very short arrays of length at most, say minSize. The program in Figure 2.15 is the generalization of the program in Figure 2.14 extended using this idea. As we have discussed, the insertion sort works poorly when the input array is sorted in descending order. Its worst-case time complexity is $O(n^2)$. Nevertheless, it runs empirically faster than does randomized quick sort when the input is very short.

Instead of sorting short intervals immediately, we can leave these intervals untouched and execute the insertion sort on the entire array only once at the end. This idea of performing one final pass is fairly efficient because the entire target array is essentially sorted, except short intervals; therefore, the insertion sort moves elements within individual short intervals. The Java program in Figure 2.16 implements this idea.

Despite efforts to improve performance, the revised quick sort might process some inputs in $O(n^2)$ time. In order to avoid the quadratic worst-case behavior of a randomized quick sort and to ensure the worst-case time complexity bound of $O(n \log_2 n)$, we can switch to $O(n \log_2 n)$-time sorting

```
public static void randomQuickSort(int[] target,
                             int aLeft, int aRight, int minSize){
    randomQuickSortSub(target, aLeft, aRight, minSize);
    insertionSort(target);  // Execute the insertion sort at the end.
}
public static void randomQuickSortSub(int[] target,
                             int aLeft, int aRight, int minSize ) {
    int left = aLeft; int right = aRight;
    while (left + minSize < right) {
        int i = partition(target, left, right);
        if( i - left <= right - i ){
            randomQuickSortSub(target, left, i-1, minSize);   left = i+1;
        }else{
            randomQuickSortSub(target, i+1, right, minSize);  right = i-1; }
    }
}
```

Fig. 2.16 A variant of the program in Figure 2.15 that executes the insertion sort on the entire target array at the end.

```
public static void introspectiveSort(int[] target,
                             int aLeft, int aRight, int minSize){
    int depthLimit = 40; // Or, =(int)(Math.log(target.length) * 0.1);
    introspectiveSortSub(target, aLeft, aRight, minSize, depthLimit);
    insertionSort(target);
}
public static void introspectiveSortSub(int[] target, int aLeft, int aRight,
                             int minSize, int depthLimit) {
    int left = aLeft; int right = aRight;
    while (left + minSize < right) {
        if(depthLimit <= 0){ // Heap sort on the range [left,right].
            heapSort(target, left, right); return; }
        depthLimit--;
        int i = partition(target, left, right);
        introspectiveSortSub(target, left, i - 1, minSize, depthLimit);
        left = i + 1;
    }
}
```

Fig. 2.17 Introspective sort.

algorithms whenever the number of recursive calls exceeds a constant, e.g., 40, or $c \log_2 n$, where c is a constant. The *introspective sort* [65] in Figure 2.17 is designed according to this idea and uses the heap sort. Most standard libraries shipped with C++/Java compilers adopt the introspective sort.

2.7 Ternary Split Quick Sort

One drawback of the randomized quick sort is that even if the pivot value occurs more than once in the target array, the algorithm generates a singleton block for one occurrence of the pivot and distributes other occurrences to the left and right blocks, which is likely to result in a situation in which the same pivot value is selected in subsequent steps. To avoid this redundancy, the ternary split quick sort [8] divides the target array into three blocks: the left block of elements less than the pivot, the middle block with all the occurrences of the pivot, and the right block of elements greater than the pivot.

Figure 2.18 illustrates the concept of the ternary split quick sort, and the program in Figure 2.19 implements it. The behavior of the algorithm is similar to that of the quick sort, but the major difference is that whenever the program finds the pivot value, it moves the pivot to the left or right end so that all occurrences of the pivot are set aside at one of the two ends. After scanning all values, the program moves the pivot values to the middle block.

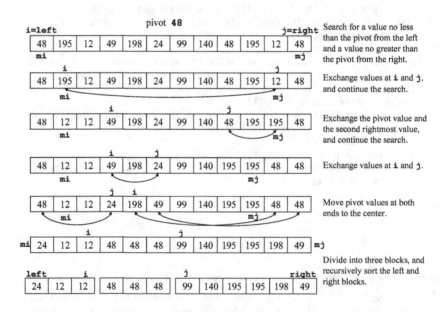

Fig. 2.18 Operation of the ternary split quick sort.

```
public static void ternarySplitQuickSort( int[] target, int left, int right ) {
    // Skip processing if the array has only one element.
    if (left >= right) return;
    // Select a pivot from the range [left,right] at random
    // using the Java Math Library.
    int pivot = target[left + (int)Math.floor(Math.random()*(right-left+1))];
    int i = left;     // i scans the array from the left.
    int mi = left;    // Position available for the next pivot value.
    int j = right;    // j scans the array from the right.
    int mj = right;   // Position available for the next pivot value.
    int tmp;
    for (;;) {
        // Move from the left until hitting a value greater than the pivot.
        for(; i <= j && target[i] <= pivot; i++)
            if(target[i] == pivot){  // Move the pivot value to the left end.
                tmp = target[i]; target[i] = target[mi]; target[mi] = tmp;
                mi++;
            }
        // Move from the right until hitting a value less than the pivot.
        for(; i <= j && pivot <= target[j]; j--)
            if(target[j] == pivot){  // Move the pivot value to the right end.
                tmp = target[j]; target[j] = target[mj]; target[mj] = tmp;
                mj--;
            }
        if (i > j)  break;
        // Exchange target[i] and target[j].
        tmp = target[i];  target[i] = target[j];  target[j] = tmp; i++; j--;
    }
    // Move the pivot values to the middle.
    for(mi--, i--; left <= mi; mi--, i--){
        tmp = target[i];  target[i] = target[mi];  target[mi] = tmp; }
    for(mj++, j++; mj <= right; mj++, j++){
        tmp = target[j];  target[j] = target[mj];  target[mj] = tmp; }
    // Sort the left and right arrays.
    ternarySplitQuickSort(target, left, i);
    ternarySplitQuickSort(target, j, right);
}
```

Fig. 2.19 Ternary split quick sort.

2.8 Radix Sort

Finally, we introduce another sorting algorithm called the *radix sort*. The radix sort treats elements in the target array as d-digits. For instance, 195 is a 3-digit, and the first digit "1" is called the *most significant digit*, while the third digit "5" is the *least significant digit*. The radix sort attempts to sort the least significant digit first, as illustrated in Figure 2.20.

The first step may shuffle elements drastically according to the order of the least significant digit. Next, it sorts elements by looking at the second least significant digit, while keeping the order of elements that have the same value in the second least significant digit. For example, in the middle of the second array of the upper picture, note that all of 195, 195, and 198 have 9 in the second least significant digit; therefore, they are moved to the third array while keeping their relative order. Then, it checks the most significant digit of these numbers to finalize their order because their order in terms of the lower two digits has been confirmed.

The last element 99 in the second array also has 9 and appears later than 195 and 199. This relative order is also conserved when 99 is moved to the third array. In the last step of processing the third array, the inconsistent order of 195, 199, and 99 is resolved by consulting the most significant digit only because their order according to the lower two digits has been settled.

The crux of the radix sort is to determine the order of integers for the focal significant digit. Those integers are typically binary or decimal. We here suppose the general assumption that the range of integers is from 1 to $m - 1$. Figure 2.21 presents how the radix sort computes the second array from the first one. It counts the number of occurrences of individual integers by scanning the target array, and it subsequently calculates the cumulative number of integer occurrences that are less than or equal to

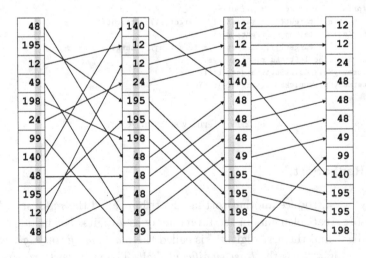

Fig. 2.20 The picture shows how the radix sort exchanges the order of 3-digits. It starts by looking at the least significant digits that are enclosed in the gray box, and subsequently continues to process higher digits.

each integer. The middle table between the two arrays displays the result. In the table, the second column presents counts of individual integers, and the third column presents cumulative counts. For example, the number of integers less than or equal to 8 (7, respectively) amounts to 10 (6), as shown in the second (third) last cell from the bottom in the third column. This information allows us to put four elements, 48, 198, 48, and 48, which have 8 in the least significant digit, into positions 6 to 9 in the second array (recall that zero-origin indexing is used).

Figure 2.22 presents a program that implements the radix sort. The second argument m indicates the number of integers in the focal significant digit. Binary and decimal digits can be processed by setting m to 2 and 10, respectively. Function digit returns the integer in the j-th least significant digit for the input element given in the first argument. In the radix sort program, array tmp is used to store temporary, partially sorted arrays, and array count is for calculating the cumulative counts of integer occurrences. Subsequently, the program scans array target from the last element, and puts all elements into the proper positions by consulting the table of cumulative counts. Whenever it fills the position in the array tmp, it decrements the corresponding count so that the count can refer to the next available position.

Finally, let us examine the computational complexity of the radix sort. Suppose that the number of elements in the target array is n, and all

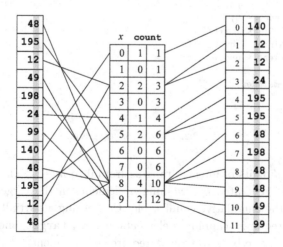

Fig. 2.21 How the radix sort computes the second array from the first one in Figure 2.20.

```
public static void radixSort( int[] target, int m ){  // target != null.
    // All elements in the array target are assumed to be nonnegative integers.
    int targetLen = target.length;
    // Array for storing a partially sorted list temporarily.
    int[] tmp = new int[targetLen];
    // Array for counting individual numbers in the significant digit.
    int[] count = new int[m];
    // Compute the maximum and set maxDigit to its number of digits.
    int max = target[0];
    for(int i=1; i<targetLen; i++){ if(max < target[i]) max = target[i]; }
    int maxDigit = 0;
    for(; max > 0; maxDigit++){ max = max/m; }
    // From the least significant digit, sort elements.
    for(int j = 1; j <= maxDigit; j++){   // j denotes the significant digit.
        // Count the number of elements equal to "i".
        for(int i=0; i<m; i++){ count[i]=0; }
        for(int i = 0; i < targetLen; i++){ count[digit(target[i], j, m)]++; }
        // Count the number of elements less than or equal to "i".
        for(int i=1; i<m; i++){ count[i] = count[i-1] + count[i]; }
        // Scan the target array from the tail, and
        // insert each element in the proper position in tmp.
        for(int i = targetLen-1; 0 <= i; i--){
            // Putting -- before count decrements count first.
            tmp[ --count[digit(target[i], j, m)] ] = target[i];
        }
        for(int i = 0; i < targetLen; i++){ target[i] = tmp[i]; }
    }
}
public static int digit(int num, int j, int m){
    int quotient = num;
    int remainder = 0;
    for(int i=0; i<j; i++){
        remainder = quotient%m;
        quotient = quotient/m;
    }
    return remainder;
}
```

Fig. 2.22 Radix sort program.

elements are d-digits of integers ranging from 0 to $m-1$. The cumulative count table is generated by scanning n elements in the target array and m counts in the cumulative count table. Computing a partially sorted list in array tmp requires scanning n elements in array target, and moving all the elements in tmp back to target requires n movements. This process is iterated d times. The total number of these basic operations is $d(3n + m)$. For example, 32-bit integers can be regarded as 32-digits of binary integers, and hence we have $d = 32$ and $m = 2$. If d and m are treated as constants, the number of basic operations of the radix sort is $32(3n + 2)$, implying that the worst-case computational complexity of the radix sort is $O(n)$.

When $m = 2$, the cumulative count table stores only two counts, which may not be effective in sorting elements. The use of a larger cumulative table accelerates the performance of the radix sort. For example, we can represent 32-bit integers using 2-digits of integers ranging from 0 to $2^{16} - 1$; namely, $d = 2$ and $m = 2^{16}$. Although the size of the cumulative table amounts to $2^{16} (= 65536)$, the number of iterations of the sorting step according to the significant digit is only 2. Therefore, the total number of operations is $2(3n + 2^{16})$, which makes sense if $n \gg 2^{16}$. Both the merge sort and randomized quick sort need approximately $n \log_2 n$ basic operations. For example, to handle a target array of one billion ($\approx 2^{30}$) 32-bit integers, the radix sort clearly outperforms the merge sort and quick sort.

Recall that in order to represent k-mer strings that consist of k nucleotides, we use k-mer integers that are $2k$-bit integers; e.g., 16-mer integers are 32-bit integers. Sorting more than one billion 16-mer nucleotide strings in a genome is common in large-scale genome processing, and the radix sort is very useful for this purpose.

Before the invention of computers, the idea of the radix sort was widely used for sorting, such as for sorting cards. Knuth [48] remarks that H. H. Seward proposed a way of implementing the idea with less memory usage in his Master's thesis in 1954.

Problems

Problem 2.1 Let $S = \{1, 2, ..., n\}$ and let f denote a permutation of S defined by a one-to-one mapping over S. Consider how to evaluate how much f preserves the ordering of S according to the number of pairs (i, j), such that $f(i) < f(j)$. Design an algorithm that computes the number in time $O(n \log_2 n)$ based on the idea of the merge sort.

Problem 2.2 Let n be a positive integer. Consider the functions $f(n)$ and $g(n)$ that are defined as follows:

$$f(n) = \begin{cases} 0 & \text{if } n = 1 \\ f(n/2) + 1 & \text{otherwise} \end{cases} \quad g(n) = \begin{cases} 0 & \text{if } n = 1 \\ 2g(n/2) + n & \text{otherwise} \end{cases}$$

Prove that $f(n) \in O(\log_2 n)$ and $g(n) \in O(n \log_2 n)$.

Problem 2.3 Give an example of length n that the median-of-3 quick sort processes in $O(n^2)$ time.

Problem 2.4 Compare the behavior of the randomized quick sort algorithm and the ternary split quick sort if all the elements in the target array are equal.

Problem 2.5 The radix sort in Section 2.8 starts swapping elements by comparing the least significant bit first. Design an algorithm that scans bits in the reverse order and begins by exchanging elements by looking at the most significant bit first.

Chapter 3

Lookup Tables

As we have seen in Chapter 1, searching a sorted list of n elements for the two boundaries of contiguous occurrences of the query takes about $2\log_2 n$ probing steps. Here, we present an efficient way of accelerating this probing operation by using two types of lookup tables called *direct-address* tables and *hash* tables. With the help of these tables, it requires a nearly constant number of steps to process one query.

3.1 Direct-Address Tables

Figure 3.1 illustrates the data structure of a direct-address table generated from intTarget* (sorted intTarget), which is used in Chapter 1. The lower portion presents a Java program for calculating direct-address tables. Let us consider how to locate the positions of ATAA, as an example. Since $encode$(ATAA) $= 48$, three occurrences of 48 and their respective positions are stored in intTarget* and posTarget as a consecutive block, which is exactly the answer to the query. A direct-address table provides a prompt method for accessing the block of the query 4-mer integer.

In more detail, the index of a direct-address table corresponds exactly to the 4-mer integer transformed from a string, and its element indicates the head of the block of the transformed integer. For example, in Figure 3.1, the 48th element of directTable, 3, indicates the head index of the block of 48 and its positions: 0, 8, and 11. Similarly, the 12th element is 0, indicating that the block of 12 starts from the 0th element in posTarget.

Indexes of directTable are prepared to represent all the possible 4-mer integers that can be transformed from strings of four nucleotide letters using $encode$, and the index ranges from 0 to $256(= 4^4)$. In Figure 3.1, eight elements in intTarget* map to directTable. How does one process

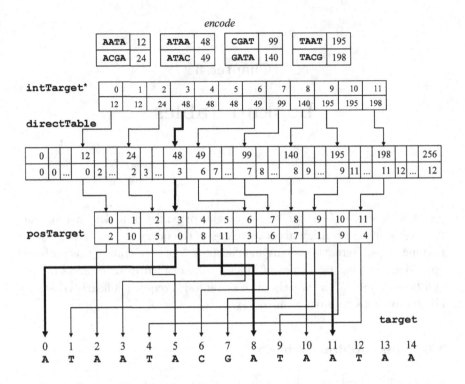

Fig. 3.1 Data structure of a direct-address table.

the other index i of directTable that does not occur in intTarget*? We set directTable[i] to min{j | i < intTarget*[j]}. This assignment may appear complex, but is useful in searching directTable.

For example, let us examine directTable to enumerate all the occurrences of 12. directTable[12] (=0) shows that the head of the block for 12 starts at the 0th position of posTarget. In addition, the next directTable[13] (=2) implies that the block terminates just before the second position of posTarget. In other words, the block of consecutive occurrences of 12 ranges from the 0th to the 1st positions in posTarget. Similarly, the block of 48 ranges from the 3rd to the 5th positions by consulting directTable[48] (=3) and directTable[49] (=6).

Figure 3.2 shows an algorithm for computing a direct-address table. It assumes that input elements in intTarget of length intTargetLen are sorted.

```
public static int[] buildDirectAddressTable( int[] intTarget, int queryLen ){
    // Assume that the elements in intTarget are sorted.
    int directTableLen = 1; // directTableLen is the queryLen-th power of 4.
    for(int i=0; i<queryLen; i++) directTableLen *= 4;
    directTableLen++; // For processing the largest element.
    // Initialize all elements in directTable to -1
    int directTable[] = new int[directTableLen];
    for(int i=0; i<directTableLen; i++) directTable[i] = -1;
    // Map all values in intTarget to directTable
    for(int i=0; i<intTarget.length; i++){
        if( !(intTarget[i] < directTableLen) ) return null;
        if(i == 0 || (0 < i && intTarget[i-1] < intTarget[i]))
            directTable[intTarget[i]] = i; }
    // Fill the remaining elements in directTable
    int nextIndex = intTarget.length;
    for(int i=directTableLen-1; 0 <= i; i--)
        if(directTable[i] == -1) directTable[i] = nextIndex;
        else nextIndex = directTable[i];
    return directTable;
}
```

Fig. 3.2 An algorithm for generating a direct-address table named `directTable`.

The direct-address table could be enormously large. To preprocess all l-mer substrings in a target string of length n, the size of `directTable` is 4^{l+1}, while the size of `posTarget` is $n-l+1$. Moreover, if the distribution of substrings in the target sequence is skewed, the direct-address table could be sparse, and only a small number of elements in `intTarget*` are actually put into `directTable`. One remedy for coping with this problem is to reduce l properly so that any element is nonempty with a high probability; however, this modification may limit the size of l to a small value. For example, let us consider the handling of the human genome sequence, which is $n = 3 \times 10^9$ in length and contains a large number of repetitive sequences. l should be less than 16 because `directTable` is smaller than `posTarget` ($4^{16} = 2^{32} \approx 4 \times 10^9$) and the number of empty elements in the direct-address table could be fairly low.

3.2 Hash Tables

Another solution for avoiding sparseness is to use a hash table instead. A hash function over a set of k-mer integers is useful for compressing a long, sparse list of fairly large numbers into a shorter list of small numbers. A hash function should be designed to eliminate collision when two distinct

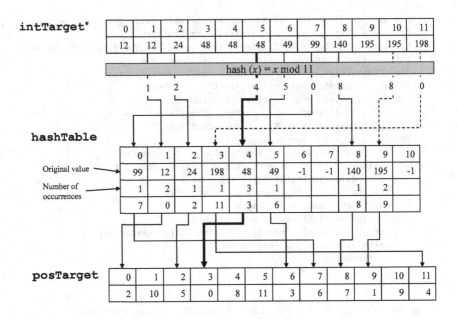

```
public static int[][] buildHashTable( int[] intTarget, int primeNumber ){
    // Create and initialize hashTable.
    int hashTable[][] = new int[primeNumber][3];
    for(int i=0; i<primeNumber; i++) hashTable[i][0] = -1;
    // Hash all elements in intTarget.
    int tLen = intTarget.length;
    for(int i=0, counter=1; i<tLen; i++, counter++){
        if(i == tLen-1 || (i < tLen-1 && intTarget[i] < intTarget[i+1])){
            // Scan hashTable until an available slot is found.
            int j = intTarget[i]%primeNumber;
            for(int k=0; hashTable[j][0] != -1; j = (j+1)%primeNumber, k++)
                if(primeNumber <= k) return null; // No empty slots.
            // Set the value, count, and start index to the empty slot.
            hashTable[j][0] = intTarget[i];
            hashTable[j][1] = counter;
            hashTable[j][2] = i-counter+1;
            counter = 0;
        }
    }
    return hashTable;
}
```

Fig. 3.3 The upper picture illustrates how the data structure of a hash table works. Below it is a Java program for generating the hash table denoted by hashTable. The program assumes that the elements in intTarget are sorted.

```
public static void queryHashTable(int[][] hashTable, int[] posTarget,
                                  int intQuery, int primeNumber){
    for(int j=intQuery%primeNumber, m=0;
        m<primeNumber && hashTable[j][0] != -1; j=(j+1)%primeNumber, m++)
        // Probe hashTable from the hash value of intTarget sequentially.
        if(hashTable[j][0] == intQuery){
            // If intQuery is found, scan posTarget to print
            // the contiguous block positions of intQuery.
            for(int k=hashTable[j][2]; k<hashTable[j][2]+hashTable[j][1]; k++)
                System.out.print(posTarget[k]+"\t");
            System.out.println();
            break;
        }
}
```

Fig. 3.4 Searching for the query using hash tables.

k-mer integers map to the same value. A common form of the hash function is hash$(x) = x$ mod p, where p is a prime number. p must be greater than the number of k-mer integers to be processed.

Figure 3.3 shows how to use a hash table instead of a direct-address table. The head values of individual blocks in `intTarget*` are mapped to indexes in `hashTable` using "hash$(x) = x$ mod 11" as a hash function. Note that 11 is selected so that it is greater than the number of distinct 4-mer integers in `intTarget*`, eight. For example, the 12 stored in the 0th element of `intTarget*` is mapped to the index 1($=$ 12 mod 11) in `hashTable`, where the value 12 and its position 0 are memorized. Similarly, 24 is mapped to index 2, where 24 and 2 are stored. In this way, we iterate the process of mapping the head value of each block.

When we attempt to map 195, although its hash value is 8, the 8th element of `hashTable` is already occupied by 140, which was given the hash value of 8 in a previous step. With such a collision, we look for an empty element in `hashTable` by checking subsequent elements one by one. We find that the 9th element is available and put the pair of 195 and its index 9 in `intTarget*` into the 9th position. Similarly, 198 maps to 0, but we see that the 0th, 1st, and 2nd elements in `hashTable` are filled. Finally, the 3rd position can accept 198. If we hit the last element of `hashTable` during a search, we restart searching from the 0th element.

Since collisions are handled in this way, care must be taken in searching for the k-mer integer representing a substring. For example, to look for TACG, we first transform TACG into 198 and calculate its hash value 0. Since the 0th element in `hashTable` is occupied by 99, we check succeeding elements one after another and find that the 3rd one contains 198. Let us consider a search for another example ACCT, which is transformed to 23 and is then mapped to 1. Starting from the 1st element, we examine subsequent elements for 23, until we hit the empty element in the 6th position. The failure to identify 23 implies that there are no occurrences of ACCT in the target sequence. Figure 3.4 presents a program that implements this idea.

This example shows that searching a hash table can be extremely time-consuming if the elements in the table are almost all occupied. To overcome this inefficiency, one can use a larger hash table by using a fairly big prime number in the hash function.

3.3 Table Size

The benefit of utilizing a hash table instead of a very sparse direct-address table is its smaller memory usage. The drawback of hashing is that all elements in the hash table may have to be examined in the worst case. However, when a reasonable hash function is used, the hashing technique is effective, and probing a few values in a hash table is usually sufficient to check the existence of the query. The construction of hash functions and in-depth analysis of their behaviors are elaborated in other books, such as [23, 48].

Another advantage of hash tables is the ability to process long substrings. In what follows, let us suppose that we must handle l-mer substrings in a target sequence of length n. If we use a direct-address table, its size could be much greater than the number of l-mer substrings, $n - l + 1$. For example, consider the human genome, which has a size of about $3 \times 10^9 (= n)$, and all the 30-mer substrings ($l = 30$); we see that $n \ll 4^{30} \approx 10^{18}$. In contrast, a hash table of roughly thrice n should be sufficient, and hence a hash table is more appropriate for processing longer substrings than a direct-address table. The real issue to solve in the hashing approach is how to sort long substrings (or their k-mer integers) efficiently before hashing integers. For this purpose, the radix sort is useful for ordering nucleotide strings.

Conversely, what happens if we must process short substrings, such that $4^l \ll n$? Since $n/4^l \gg 1$, the direct-address table is not sparse, but is densely occupied by elements. For example, in the human genome, a 10-mer substring has too many occurrences because $(3 \times 10^9)/4^{10} \approx 3 \times 10^3$. Consider the cases when all 4^l l-mer substrings appear in the target sequence. The size of the direct-address table is 4^{l+1}, while the size of a hash table is roughly twice or thrice 4^l because all 4^l l-mer substrings occur in the target sequence. Moreover, a hash table imposes the extra cost of calculating its hash function. Therefore, when one substring has many occurrences on average, a direct-address table is more efficient than a hash table.

Table 3.1 summarizes the sizes of direct-address and hash tables. The sizes are calculated by assuming that n is representable in 32 bits (4 bytes), i.e., $2^{24} < n \leq 2^{32}$. This assumption is feasible because the sizes of currently available vertebrate genomes are less than $3 \times 10^9 (< 2^{32} \approx 4 \times 10^9)$. For example, the size of a direct-address table for $l = 15$ and $n = 3 \cdot 10^9$ approximates $4 \cdot (4^l + n)$, which is about 16 gigabytes (16×10^9).

The exact size of a hashTable is difficult to define because the precise number of entries (blocks), which are bounded by $\min(n - l + 1, 4^l)$, is difficult to predict due to duplicate substrings. Another reason is that we can change the size of a hash table flexibly by solving the trade-off between increasing the efficiency of probing a larger hash table and reducing the hash table size. In Table 3.1, α indicates the size of the hashTable in terms of the number of elements put in the table, and α is usually between 2 and 4.

Table 3.1 Sizes are expressed in bytes. The number of elements in posTable is $n - l + 1$, which nearly equals n, since $l \ll n$. The size of a hash table is dependent on a constant factor α, which is between 2 and 4 in practice.

	direct-address table	hash table
directTable	$4 \cdot (4^l + 1)$	-
hashTable	-	$< 12\alpha \cdot \min(n, 4^l)$
posTable	$\approx 4n$	$\approx 4n$
Total	$\approx 4 \cdot (4^l + n)$	$< 12\alpha \cdot \min(n, 4^l) + 4n$

3.4 Using the Frequencies of k-mers

We have seen that a direct-address table is more space efficient than a hash table when l is sufficiently small, so that 4^l is much smaller than n. If $n = 3 \times 10^9$ and $l = 12$, we have $4^{12} \approx 4 \times 10^6 \ll 10^9$. In these cases, to build a direct-address table, instead of using traditional sorting algorithms, one can use the more efficient counting algorithm used in the radix sort.

Figure 3.5 illustrates this idea. Here, we use **AACAACACAA** as the sample target sequence instead of the running example used thus far. Let us transform 2-mer substrings into 2-mer integers, e.g., *encode*(**AA**) = 0, *encode*(**AC**) = 1, and *encode*(**CA**) = 4. First, we build `intTarget` from this target sequence, as we have done before. In `intTarget`, only three 2-mer integers appear (0, 1, and 4), and individual integers have multiple occurrences. In order to sort such a list of 2-mer integers where a fairly small number of 2-mer integers occur many times, rather than using the quick, merge, or radix sort algorithms, it is better to build the frequency table of individual 2-mer integers by scanning the elements in `intTarget` only once to count the number of each occurrence. The frequency table helps us to generate `intTarget*` because the table tells us that the first three elements in `intTarget*` should contain 0, the next three elements contain 1, and the last three contain 4.

In order to build `posTable`, we scan `intTarget` from right to left and need to put the index (position) of each k-mer integer into the proper slot in `posTable`. For this purpose, it is useful to have a cumulation table in which each k-mer integer is associated with the sum of frequencies of those k-mer integers that are lower than or equal to the focal k-mer integer. In the cumulation table, the value for each k-mer integer indicates the available slot into which the next position should be placed. For example, we put the index of the last element, 0, in `intTarget` into the 2nd slot of `posTable`, the penultimate element, 4, into the 8th, and the third last, 1, into the 5th. During the computation, every time we hit an occurrence of 4, we decrement the value for 4 in the cumulation table to memorize the next empty slot.

Overall, `intTarget` is scanned only twice to create a direct-address table. If one wonders why we can avoid sorting, it is because all the k-mer integers are sorted implicitly when frequencies are counted. Therefore, this algorithm is efficient as long as the number of distinct k-mer integers in the list is fairly small, so that the table fits into the main memory. Otherwise, a huge frequency table may be generated.

3.5 Techniques for Reducing Table Size

Here, we introduce two ways for reducing the size of lookup tables. One approach considers a partial set of substrings in place of the entire set. For example, in the target string **ATAATACGATAATAA**, select substrings of length

Lookup Tables

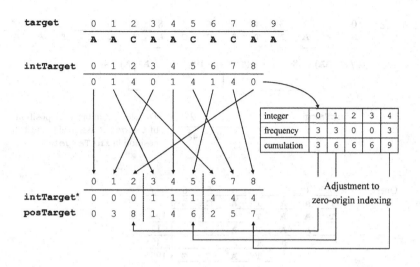

```
public static void sortIntTargetByCounting(
                int[] intTarget, int posTarget[], int queryLen ) {
    // Assume that the input intTarget is not sorted yet.
    // Set variable tableLen to the queryLen-th power of 4
    int tableLen = 1;
    for(int i=0; i<queryLen; i++) tableLen *= 4;
    // Count the frequency and the cumulation for each integer representation.
    int[] frequency = new int[tableLen];
    for(int i=0; i<tableLen; i++)            frequency[i] = 0;
    int intTargetLen = intTarget.length;
    for(int i=0; i<intTargetLen; i++) // Exit if intTarget[i] >= tableLen.
        if(intTarget[i] < tableLen) frequency[intTarget[i]]++; else return;
    int[] cumulation = frequency;   // Reuse the frequency table space.
    for(int i=1; i<tableLen; i++)
        cumulation[i] = cumulation[i-1] + frequency[i];
    // Adjust to zero-origin indexing.
    for(int i=0; i<tableLen; i++) cumulation[i]--;
    // Generate intTargetSort for storing the sorted intTarget.
    int intTargetSorted[] = new int[intTargetLen];
    for(int i = intTargetLen-1; 0 <= i; i--){
        intTargetSorted[cumulation[intTarget[i]]] = intTarget[i];
        posTarget[cumulation[intTarget[i]]] = i;
        cumulation[intTarget[i]]--;
    }
    for(int i=0; i<intTargetLen; i++)
        intTarget[i] = intTargetSorted[i];
}
```

Fig. 3.5 The picture illustrates how an efficient algorithm for generating a direct-address table works, and a Java program for it is shown below it.

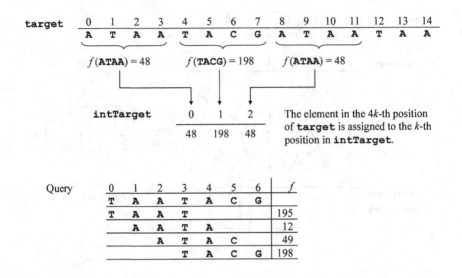

Fig. 3.6 Nonoverlapping substrings are stored to reduce the size of a lookup table.

four that start at the $4n$-th position for $n = 0, 1, 2$, so that they do not overlap, and put the k-mer integer of each nonoverlapping substring into the n-th (not $4n$-th) position of intTarget to reduce its size. See Figure 3.6. The use of *nonoverlapping* substrings dramatically reduces the size of a lookup table at the sacrifice of failing to return correct answers to some queries. For example, the 4-mer integer transformed from TAAT is 195, which is missing in intTarget. Although 48 for ATAA can be found in intTarget, its occurrence in the 11th position disappears in intTarget, which is also misleading. Despite these difficulties, keeping nonoverlapping substrings is still useful in most applications.

For example, let us scan the target sequence for seeking the query TAATACG. Since intTarget has 4-mer integers of 4-mer substrings, probe intTarget for 4-mer integers of individual 4-mer substrings in TAATACG from left to right. Although the 4-mer integer 195 for TAAT, 12 for AATA, and 49 for ATAC are not found, 198 for TACG is identified, and hence we move on to aligning the remaining three letters TAA to the target to find that TAATACG fully matches a substring in the target. This approach is not always successful because making a query of length 6 may result in a failure, e.g., consider TAATAC.

Another table size-reduction technique divides the target sequence into partitions of length 2^{16} and subsequently generates a lookup table for each partition so that k-mer integers can be represented in 2 instead of 4 bytes, which halves the size of posTarget. This method, however, must be applied with care. For example, if the target is of length $n = 3 \times 10^9$, the method generates 45,777 partitions and the same number of lookup tables for individual partitions. If 8-mer substrings are considered, the number of different 8-mer substrings is $4^8 (= 2^{16})$, and hence each substring is likely to appear once in one partition, the length of which is 2^{16}.

However, if 13-mer substrings are preprocessed, each 13-mer substring occurs in one partition with a probability of $2^{16}/4^{13} = 1/1024$, indicating the inefficiency of scanning all the partitions since the identification of the substring in one partition is successful in only one case out of 1,024 trials checking partitions. In this case, in order to gain efficiency, we should use a 4-byte representation for k-mer integers without reducing the size of the lookup table.

Problems

Problem 3.1 Consider the hash function "hash$(x) = x$ mod p," and discuss difficulties when p is not a prime number by giving an example.

Problem 3.2 Consider the set of distinct 4-mer integers of intTarget* in Figure 3.3, {12, 24, 48, 49, 99, 140, 195, 198}. What is the minimum prime number p such that the hash function "hash$(x) = x$ mod p" maps all the 4-mer integers in the set to different values?

Problem 3.3 In Table 3.1, suppose that nucleotide letters occur at random in a target sequence of length $n = 3 \times 10^9$, and let $\alpha = 2$.

(1) Let $l = 15$. Roughly estimate the size of the hash table.
(2) What is the value of l such that a hash table is smaller than a direct-address table?

Problem 3.4 The direct-address and hash tables described in this chapter are designed by assuming that their input intTarget is static and is not modified. Therefore, these data structures are not suitable for handling updates in the input array. Consider how to update these lookup tables in order to process requests to insert one element into the input array or delete one from the input.

Chapter 4

Suffix Arrays

Lookup tables such as direct-address tables and hash tables are designed to access queries of fixed length in a nearly constant time. However, in order to handle queries of various lengths by using lookup tables of l-mer strings, long queries must be partitioned and preprocessed in some way, e.g., some l-mer substrings in a query are selected and searched. In this chapter, we introduce *suffix trees* and *suffix arrays*, which are data structures for manipulating queries of various lengths in a straightforward manner.

Definition 4.1 Let S denote the target string $b_0 b_1 \ldots b_{n-1}$ of length n. The i-th element of S is described by $S[i]$. The substring of S that ranges from the l-th position to the r-th position, $b_l, \ldots, b_r$, is denoted by $S[l, r]$, where $l, r \in [0, n-1]$. A *prefix* is a substring starting from the 0-th position, $S[0, r]$, while a *suffix* is a substring ending at the last position, $S[l, n-1]$.

S	b_0	...	b_l	...	b_r	... b_{n-1}
$S[0, r]$	b_0	...	b_l	...	b_r	
$S[l, n-1]$			b_l	...	b_r	... b_{n-1}

Prefixes and suffixes are *proper* if they are shorter than the original string.

Example 4.1 Let S denote ATAATACGATAATAA. In the following table, the left half shows proper prefixes of S, while the right presents proper suffixes of S.

$S[0,0]$	=	A	$S[10,14]$	=	AATAA
$S[0,1]$	=	AT	$S[11,14]$	=	ATAA
$S[0,2]$	=	ATA	$S[12,14]$	=	TAA
$S[0,3]$	=	ATAA	$S[13,14]$	=	AA
$S[0,4]$	=	ATAAT	$S[14,14]$	=	A

4.1 Suffix Trees

A suffix tree is a tree data structure that organizes all suffixes in the input string.

Definition 4.2 Let S be a string of n characters such that its end character is $, and $ does not appear anywhere else. A *suffix tree* T for S is a rooted tree that meets the following conditions:

(1) T has n leaf nodes.
(2) An individual internal node has multiple child nodes. Edges from an internal node to children are labeled with nonempty substrings of S that have different initial characters.
(3) Concatenating substrings labeled on edges from the root to a leaf node gives a suffix that starts at position k for some k. The leaf node is labeled with k. For any k $(0 \leq k \leq n-1)$, there is a unique leaf node labeled with k.

Figure 4.1 illustrates the suffix tree for ATAATACGATAATAA$. Observe that concatenating substrings on the path from the root to leaf, e.g., 8, gives ATAATAA$, which is equal to the suffix starting at the 8th position.

The suffix tree is useful in enumerating positions where the given query appears. Consider searching the suffix tree in Figure 4.1 for ATAA. From the root, the following edges labeled A, TA, and A sequentially reach the root of the subtree with leaf nodes labeled 0, 8, and 11. The absence of nonexistent strings is also easy to check, e.g., searching for AAG scans two edges labeled with A but finds no edge with a string starting with G.

In the suffix tree, edges labeled with lexicographically lower strings are shown in the upper part. Traversing nodes in the suffix tree by selecting upper edges and visiting their child nodes first allows us to scan leaf nodes labeled with suffixes in lexicographic order. It is easy to seek multiple query strings that are sorted in lexicographic order by traversing the suffix tree according to the strategy of selecting upper edges first.

It is crucial to add the character $ that does not appear anywhere else to the end of the string. Otherwise, the removal of $ eliminates the leaf node labeled with 13 that represents the suffix starting at position 13, AA. This elimination makes the tree violate the third requirement for suffix trees; i.e., there must be a unique leaf node labeled with each position.

Weiner proposed the idea of suffix trees in 1973 [104]. McCreight in 1976 reported an efficient but complex algorithm that builds a suffix tree

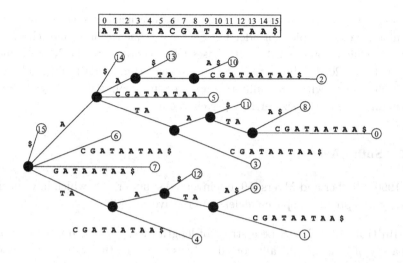

Fig. 4.1 Suffix tree for ATAATACGATAATAA$.

in a time proportional to the length of the input string [63]. Ukkonen gave another simpler linear-time algorithm for this purpose in 1995 [102]. For more details of suffix tree construction, see [35].

Although suffix trees are fundamental to string processing, they are not widely used in practical software programs because they involve extensive space usage. A variety of data representations for suffix trees have been proposed. Kurtz presented a space-efficient representation that requires $20n$ bytes in the worst case (where n is the input length) and $10.1n$ bytes on average for many data sets of different types when integers are expressed using 4 bytes [52]. The space requirement involves a temporary space for calculating a suffix tree in addition to the space for storing the suffix tree itself. It is difficult to estimate the precise space size for building a suffix tree because its size is largely dependent of the characteristics of the data.

In the following sections, we introduce the notion of a suffix array. A variety of suffix array construction algorithms have been proposed. Constructing the suffix array of an input empirically needs less space and runs faster than making the suffix tree for the same input. For instance, the Larsson-Sadakane algorithm, which will be explained in this chapter, requires $8n$ bytes. One might think that the difference in the space requirements between suffix trees and arrays is fairly small. Suffix trees and arrays should be stored in expensive main memory for efficiency because data in these structures are accessed at random. Moreover, the data structures often demand a large amount of space, making it crucial to save costly main memory. For example, consider the genomic sequence of human chromosome 1, which is about 2.46×10^8 bases long according to the NCBI Build 35 as of May 2004. The Larsson-Sadakane algorithm needs 1.968 gigabytes ($= 1.968 \times 10^9$), while the suffix tree may require about 2.48 gigabytes even assuming an average size of 10.1 per character.

4.2 Suffix Arrays

In 1990, Manber and Myers [59, 60] invented suffix arrays, which have been widely accepted as a space-efficient alternative to suffix trees.

Definition 4.3 Let S be a string of length n. A *suffix array* SA of S is an array of lexicographically sorted suffixes of S such that $SA[i] = k$ if and only if the i-th suffix in the lexicographic order starts at position k in S. An *inverse suffix array* ISA is such an array that $ISA[k] = i$ if and only if $SA[i] = k$. In other words, the suffix starting at k position in S has rank $ISA[k]$ in lexicographic order. We assume that both SA and ISA have zero-origin indexing.

Figure 4.2 presents a suffix array and its inverse. We assume that $ is lower than other letters. The figure conceptually displays all suffixes. Figure 4.3 shows the one-to-one correspondence between leaf nodes in the suffix tree and elements in the suffix array. If we traverse the suffix tree by selecting edges labeled with lexicographically lower strings first, we visit leaf nodes in the lexicographical order of their suffixes that coincides with the order of the suffixes in the suffix array.

For a very long input string, suffixes are likely to be numerous and long. We must therefore avoid generating all suffixes when we build suffix arrays. We will present space-efficient algorithms for creating them.

Before the description, we discuss how to search suffix arrays for a query of interest. Searching the suffix tree for the query is straightforward because one can simply compare characters in the query with those in strings of edges, select the matching edge, move to the child node, and iterate the same step. However, suffix arrays do not involve explicit counterparts of internal nodes in suffix trees, making it a nontrivial question to search the suffix array efficiently for the query. In what follows, we present an efficient way of simulating the search of suffix trees using suffix arrays and supplementary information.

Suffix Arrays

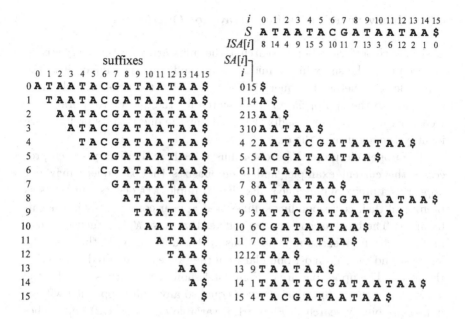

Fig. 4.2 The suffix array and the inverse suffix array of ATAATACGATAATAA$.

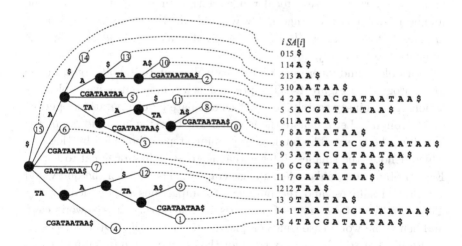

Fig. 4.3 Correspondence between leaf nodes in the suffix tree and elements in the suffix array of ATAATACGATAATAA$.

4.3 Binary Search of Suffix Arrays for Queries

Basically we perform a binary search of the suffix array for the query string; i.e., we probe the suffix in the middle of the suffix array, compare it to the query, decide whether the query is lower or higher than the probed suffix, and move to the appropriate half to search for the query. We iterate this step until we encounter a suffix that contains the query as its prefix or we identify the absence of the query.

The upper picture in Figure 4.4 illustrates how the binary search processes the current example, and it emphasizes that the query may not appear as a prefix of only one suffix. For example, ATA occurs four times as prefixes of their corresponding suffixes that appear contiguously in the suffix array. The block of consecutive suffixes that have ATA as their common prefix ranges from six to nine positions in the suffix array. In this case, the leftmost and the rightmost boundaries of the range are required to describe the block. The upper picture of Figure 4.4 shows the binary search for the leftmost boundary. Since no query is equal to any suffix appended with $, unlike the binary search in Figure 1.5, variable l_i ($i = 0, 1, 2$) approaches the leftmost boundary from the lowest element and variable r_i ($i = 0, 1, 2$) from the highest, so that the suffix at l_i is lexicographically lower than the query, the suffix at r_i is higher than the query, and r_i reaches the leftmost boundary. Arcs represent focal ranges during the computation, and the middle position in the range is always probed. For example, the longest arc stands for the range $[l_0, r_0]$, and the suffix at its middle position 7 is compared to the query.

Variable l_i and r_i present temporary values of left and right in the lower program of Figure 4.4. The program is a simple but naive implementation of the binary search idea. Let us estimate its computational performance roughly. Let n denote the length of the target string, and m_q be the length of the query. The number of probing steps is bounded by $\lceil \log_2 n \rceil$, which denotes the smallest integer that is greater than or equal to $\log_2 n$. Each probing step makes elementwise comparisons between the query and the probed suffixes, and the number of comparisons is bounded by m_q. Therefore, the total number of comparisons is $m_q \lceil \log_2 n \rceil$ in the worst case, and hence the worst-case time complexity of the search is $O(m_q \log_2 n)$.

Recall that searching suffix trees for the query visits internal nodes in a straightforward manner and its computation time is $O(m_q)$. We will further improve the simple binary search so that suffix arrays become a competent alternative to suffix trees in terms of search efficiency.

Suffix Arrays

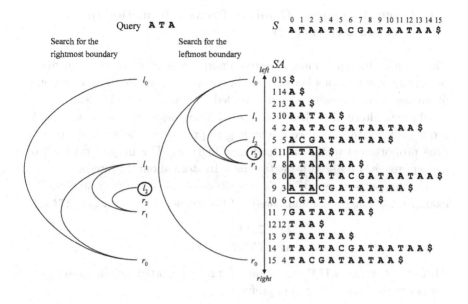

```
public static int searchLeftmost(int[] S, int[] SA, int[] query){
    // Exit if the query, which is greater than "$", is outside the scope.
    if(compare(S,SA[SA.length-1], query) == -1) return -1;
    int left, right;
    for(left = 0, right = SA.length-1; left+1 < right; ){
        int middle = (left + right) / 2;
        if(compare(S, SA[middle], query) == -1) left = middle;
        else right = middle; }
    if( compare(S, SA[right], query) == 0 ) return right;
    else return -1; // No occurrences of the query.
}
public static int compare(int[] S, int start, int[] query){
    // Return 0 if the query occurs as a prefix of the suffix S[start].
    // Otherwise, return -1 and 1 if the suffix from "start" is
    // lexicographically lower or higher than query.
    for(int i=0; start + i < S.length && i < query.length; i++)
        if(S[start+i] < query[i]) return -1;
        else if(S[start+i] > query[i]) return 1;
    return 0;
}
```

Fig. 4.4 Simple, but naive, binary search of suffix arrays. The upper picture shows the suffix array SA vertically, and suffixes are displayed horizontally. Recall that in the binary search presented in Figure 1.5 of Section 1.5, the horizontal array is searched for the leftmost and rightmost boundaries of contiguous occurrences of the query. To use the same wording, the upper and lower directions are treated as the left and right.

4.4 Using the Longest Common Prefix Information to Accelerate the Search

The simple but naive binary search algorithm is likely to perform many redundant comparisons because it always starts from the initial elements of suffixes, and this can be safely avoided in typical cases. We will present an efficient alternative that takes $O(m_q + \log_2 n)$ comparisons to handle a query of length m_q, if we are willing to preprocess the suffix array in a time proportional to the target string length n. The major drawback of this approach is its need for extra space to store about $2n$ integers.

For this purpose, we utilize the *longest common prefix* of two strings. For example, the longest common prefix of the following two strings is **ATAATA**.

ATAATAA$
ATAATACGATAATAA$

Although **ATA** and **ATAAT** are common prefixes shared by the above two, neither is the longest common prefix.

Definition 4.4 Let S denote the target string, let n be the length of S, and let SA be the suffix array of S. The length of the longest common prefix of suffixes $S[SA[i], n-1]$ and $S[SA[j], n-1]$ is denoted by $lcp(i,j)$ for simplicity. The length of the longest common prefix of $S[SA[i], n-1]$ and the query string $query$ is designated by $lcp(i, query)$.

The definition of the symbol for the longest common prefix is restricted so that the two arguments must be elements in the suffix array or the query.

We assume that the query exists between the suffix starting at l and the suffix at r in terms of lexicographic order, as illustrated in Figure 4.5. The binary search algorithm probes the middle suffix starting from $m = (l+r)/2$ to check the lexicographic order between the query and the middle suffix in order to decide which half of the range to search next for the query. The order can be determined by a single elementwise comparison if $lcp(m, query)$ is available because both strings share the longest common prefix of that length, and it is sufficient to examine two elements at position $lcp(m, query)$; these two elements must be different due to choosing of the longest common prefix.

We compute $lcp(m, query)$ by using the inductive hypothesis that the lengths of the longest common prefixes between the query and both ends, $lcp(l, query)$ and $lcp(r, query)$, have been calculated in previous steps. In addition, we assume that $lcp(l, r)$, $lcp(l, m)$, and $lcp(m, r)$ are also available, although it is nontrivial to design an efficient algorithm for computing these values, as will be explained later.

The left and right columns in Figure 4.5 show two conditions that are applicable: when $lcp(m,r) \leq lcp(l,m)$ and when this condition is not true. Since we can handle these two cases similarly, we will explain the four patterns in the left column from top to bottom.

- If $lcp(l, query) < lcp(l, m)$, the query and middle suffix must have the longest common prefix of length $lcp(l, query)$, so we see that $lcp(m, query)$ equals $lcp(l, query)$ without making any elementwise comparisons. Since we assume that the query is lexicographically equal to or higher than the left suffix starting at l, the query must be lexicographically higher than the middle suffix, and we set l to m.
- If $lcp(l, m) < lcp(l, query)$, then $lcp(m, query)$ must be equal to $lcp(l, m)$, and the query is lower than the middle suffix.
- If $lcp(l, m) = lcp(l, query)$, then $lcp(l, m) \leq lcp(m, query)$, and we calculate $lcp(m, query)$ by comparing characters in the query and middle suffix from position $lcp(l, m)$. Subsequently, comparing two elements at position $lcp(m, query)$ determines the order of the query and the middle suffix.

In the binary search, at most $\lceil \log_2 n \rceil$ elements in the suffix array are probed. Each probing step performs two comparisons between lcp values. During the binary search, in the third condition, elements in the query are compared with those in the middle suffix at most $(m_q + \lceil \log_2 n \rceil)$ times, where m_q is the length of the query. Therefore, the total number of comparisons is $O(m_q + \log_2 n)$.

Figures 4.6 presents a program code that implements the above idea. The program assumes that $lcp(j-1, j)$ for neighboring suffixes in the suffix array is available in LCP[j], and $lcp(i, j)$ is in LCP_AUX[(i+j)/2] for ranges (i, j) $(2 \leq j - i)$ that the binary search may examine. One might be concerned that two different ranges (i, j) might generate the same value $(i+j)/2$, where $(i+j)/2$ is the integer quotient. However, this conflict does not happen for ranges that the binary search might check. This is because if the binary search divides the range (i, j) into two subranges $(i, (i+j)/2)$ and $((i+j)/2, j)$ that are of size more than one, it is obvious that their middle points are different from $(i+j)/2$. Figure 4.7 illustrates such an example. In the next subsection, we will present an efficient algorithm for building these lcp arrays.

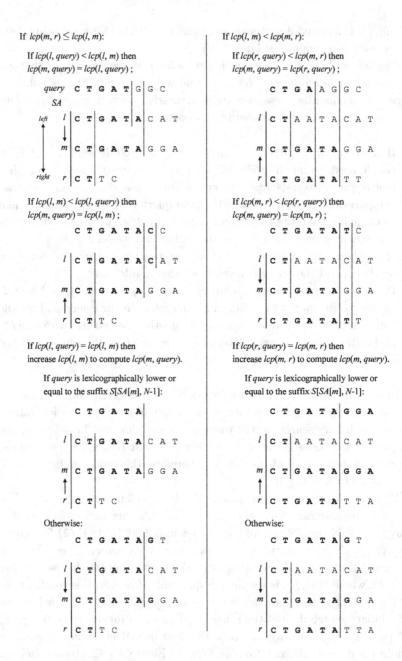

Fig. 4.5 Accelerating the binary search of suffix arrays with lcp information. Contiguous bold characters indicate the longest common prefixes of string pairs.

```
public static int searchLeftmost(int[] S, int[] SA, int[] query,
                    int[] LCP, int[] LCP_AUX){
    // Compute the lcp values between the query and both end suffixes.
    int left = 0;
    int lcp_left_query = 0;
    int right = SA.length-1;
    int lcp_right_query = lcp0(0, query, S, SA[right]);
    // Exit if the query, which is greater than "$", is outside the scope.
    if(!less_eq(lcp_right_query, query, S, SA[right])) return -1;
    // Binary search.
    for(int middle=(left+right)/2; left+1 < right; middle=(left+right)/2){
            if(   lcp1(left, middle, LCP, LCP_AUX)
               >= lcp1(middle, right, LCP, LCP_AUX)){
                if(lcp_left_query < lcp1(left, middle, LCP, LCP_AUX))
                    left = middle; // lcp_left_query remains unchanged.
                else if(lcp_left_query > lcp1(left, middle, LCP, LCP_AUX)){
                    right = middle;
                    lcp_right_query = lcp1(left, middle, LCP, LCP_AUX);
                }else{  // Set the lcp of the query and the middle suffix to i.
                    int i = lcp0( lcp1(left, middle, LCP, LCP_AUX),
                            query, S, SA[middle]);
                    if(less_eq(i, query, S, SA[middle])){
                        right = middle; lcp_right_query = i;
                    }else{ left = middle; lcp_left_query = i;} }
            }else{
                if(lcp_right_query < lcp1(middle, right, LCP, LCP_AUX)){
                    right = middle; // lcp_right_query remains unchanged.
                }else if(lcp_right_query > lcp1(middle, right, LCP, LCP_AUX)){
                    left = middle;
                    lcp_left_query = lcp1(middle, right, LCP, LCP_AUX);
                }else{  // Set the lcp of the query and the middle suffix to i.
                    int i = lcp0( lcp1(middle, right, LCP, LCP_AUX),
                            query, S, SA[middle]);
                    if(less_eq(i, query, S, SA[middle])){
                        right = middle; lcp_right_query = i;
                    }else{ left = middle; lcp_left_query = i;}
    }   }   }
    if(lcp_right_query == query.length) return right; else return -1;
}
public static boolean less_eq(int lcp, int[] query, int[] S, int start){
    // Return true if the query is lower than or equal to the suffix.
    if(lcp == query.length) return true;
    if(start+lcp < S.length && query[lcp] < S[start+lcp]) return true;
    return false;  }
public static int lcp0(int offset, int[] query, int[] S, int start){
    int i = offset; // Compute lcp.
    for(; i < query.length && start+i < S.length && query[i]==S[start+i]; i++);
    return i;  }
public static int lcp1(int i, int j, int[] LCP, int[] LCP_AUX){ // Scan tables.
    if(i+1 == j) return LCP[j]; else return LCP_AUX[(i+j)/2];  }
```

Fig. 4.6 Program for accelerating the binary search of suffix arrays with lcp information.

4.5 Computing the Longest Common Prefixes

If $lcp(k-1, k)$ is available for any k, computing $lcp(i, j)$ for $2 \leq j - i$ is straightforward, as it is the minimum value of $lcp(k-1, k)$ for $k = i+1, \ldots, j$, which can be proved by the induction on $j - i$.

Proposition 4.1 $lcp(i, j) = \min\{lcp(k-1, k) \mid k = i+1, \ldots, j\}$ for $2 \leq j - i$.

We do not have to compute $lcp(i, j)$ for all possible ranges beforehand, but need to consider ranges that the binary search may examine. These ranges are illustrated by arcs in Figure 4.7. Note that the number of ranges to preprocess is only $2n - 3$, where n is the size of SA and the length of the target string. More precisely, there are $n - 1$ ranges of size one and $n - 2$ ranges of size more than one. Integers labeled with individual arcs present the lengths of longest common prefixes that can be calculated in a bottom-up manner. For example, $lcp(1, 3) = \min\{lcp(1, 2), lcp(2, 3)\}$ and $lcp(0, 3) = \min\{lcp(0, 1), lcp(1, 3)\}$. Repeating this process computes lcp values in $O(n)$ time for all ranges of size more than one that the binary search may check. The computation of $lcp(k-1, k)$ remains to be described. Kasai, Lee, Arimura, Arikawa, and Park proved the following theorem [45].

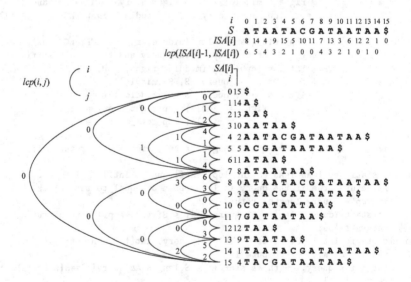

Fig. 4.7 Preprocess a suffix array to calculate the lcp values for all possible pairs of suffixes that may be probed during the binary search.

Theorem 4.1 For $i = 0, 1, \ldots, n-3$,

$$lcp(ISA[i] - 1, ISA[i]) - 1 \leq lcp(ISA[i+1] - 1, ISA[i+1])$$

Proof. The proof is trivial if $lcp(ISA[i] - 1, ISA[i]) = 0$. Otherwise, removing the first elements from the two suffixes starting at $SA[ISA[i] - 1]$ and $i (= SA[ISA[i]])$ yields two suffixes starting at $SA[ISA[i] - 1] + 1$ (denote this by k) and $i + 1$. Since the longest common factor of the former two suffixes is longer by one than that of the latter two,

$$lcp(ISA[i] - 1, ISA[i]) - 1 = lcp(ISA[k], ISA[i+1]).$$

Although $ISA[k] < ISA[i+1]$, they are not always consecutive; e.g., observe that when $i = 2$, we have $k = SA[ISA[2] - 1] + 1 = 11$, $ISA[k] = 6$ and $ISA[i+1] = 9$. Figure 4.8 illustrates this case by depicting the relevant leaf nodes in the suffix array. Finally, according to Proposition 4.1, we have

$$lcp(ISA[k], ISA[i+1]) \leq lcp(ISA[i+1] - 1, ISA[i+1]). \qquad \square$$

Figure 4.7 shows values of $lcp(ISA[i] - 1, ISA[i])$ for $i = 0, 1, \ldots, 13$; observe that the above inequality holds for $i = 0, 1, \ldots, 13$.

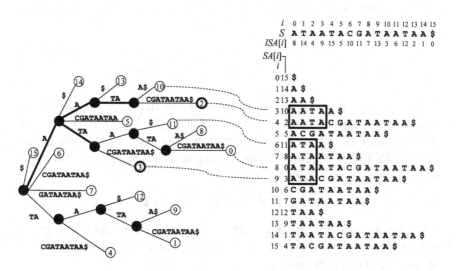

Fig. 4.8 The suffixes starting at positions 2 and 3 are highlighted by bold circles in the left suffix tree. The bold branches illustrate the two longest common prefixes, one is between $10 (= SA[3])$ and $2 (= SA[4])$, and the other between $11 (= SA[6])$ and $3 (= SA[9])$.

```
public static void buildLcp(int[] S, int[] SA, int[] LCP, int[] LCP_AUX){
    // Assume that the last symbol of S is "$", and SA is the suffix array of S.
    // int[] LCP = new int[S.length]; int[] LCP_AUX = new int[S.length-1];
    int n = SA.length;
    int[] ISA = new int[n];
    for(int i=0; i<n; i++) ISA[SA[i]]=i; // Build the inverse suffix array.
    for(int lcp=0, i=0; i<n; i++) {
        int k = ISA[i];          // Suffixes are processed from longer ones.
        if(k==0) LCP[k] = -1;    // The last suffix "$".
        else {
            int j = SA[k-1];     // j=SA[k-1] and i=SA[k] since k=ISA[i].
            while(j+lcp<=n && i+lcp<=n && S[j+lcp]==S[i+lcp]) lcp++;
            LCP[k] = lcp; }
        // if(lcp == 0) lcp++; can be added for ease of understanding.
        if(lcp > 0) lcp--;
    }
    // Compute lcp values for all possible intervals and put them into
    // LCP_AUX according to LCP_AUX[(l+r)/2] = lcp(l,r) if l+1 < r.
    buildLcp1(0, SA.length-1, LCP, LCP_AUX);
}
public static int buildLcp1(int l, int r, int[] LCP, int[] LCP_AUX){
    if(l+1 == r) return LCP[r];
    else{ int v = Math.min(buildLcp1(l, (l+r)/2, LCP, LCP_AUX),
                           buildLcp1((l+r)/2, r, LCP, LCP_AUX));
        LCP_AUX[(l+r)/2] = v;
        return v;
    }
}
```

Fig. 4.9 Program for computing $lcp(i,j)$.

The above theorem allows us to compute the right side by making the elementwise comparison at the offset position $lcp(ISA[i]-1, ISA[i]) - 1$ on the left side. Figure 4.9 shows a program that implements this idea. Recall that $lcp(ISA[i]-1, ISA[i])$ denotes the length of the longest common prefix of two neighboring suffixes

$$S[SA[ISA[i]-1], n-1] \text{ and } S[SA[ISA[i]], n-1]$$

in the suffix array. The program performs elementwise comparison between these two suffixes by S[j+lcp] == S[i+lcp], which is the dominant factor of the overall computation.

We estimate the total number of elementwise comparisons. To this end, it is helpful to add an extra statement if(lcp == 0) lcp++; before if(lcp > 0) lcp--;. Note that the extra statement performs increments lcp only when lcp is zero, while the next statement immediately produces decrements lcp. Therefore, the virtual statement has no side effects, but it ensures that the program always performs lcp++ or lcp-- after making one comparison. Therefore, the total number of comparisons is bounded by the total number of these increments and decrements. One decrement is performed for each value of i, indicating that the total number of decrements is n. lcp is initially set to zero and must be less than n when the program terminates, because lcp is the length of the longest common prefix of two suffixes in the input. Since the final value of lcp indicates the number of increments minus the number of decrements, the total number of increments must be less than $2n$. We therefore see that the number of comparisons is bounded by $3n$. When $lcp(k-1, k)$ is available, we calculate $lcp(i, j)$ for $n-2$ ranges by executing buildLcp1 recursively. Consequently, the worst case time complexity of buildLcp is $O(n)$.

4.5.1 *Application to Occurrence Frequencies of k-mers*

We have seen that the longest common prefix information is useful for improving query performance. In reality, the lcp information has been widely used in many other applications, and we here briefly mention two cases relating to large-scale genome processing.

Design of short nucleotide k-mer strings such that k ranges from 20 to 60 is essential to observe or knockout a specific gene of interest without cross-reacting with off-target genes. The technique is widely used in designing primers for polymerase chain reaction (PCR), oligonucleotide probes for DNA microarrays, and siRNA sequences for RNA interference. We here introduce some concepts relating to measuring the uniqueness of k-mer strings.

The simplest measure is the number of occurrences of a k-mer string in the target sequence, which is called *occurrence frequency*. Each k-mer string appears as prefixes of some suffixes that are sorted in the suffix array. Thus, each k-mer must appear consecutively in the suffix array, and the length of the run of the k-mer equals its occurrence frequency. In order to implement this method, consider a k-mer that appears first at position $SA[i]$ in the suffix array. Observe that the k-mer also appears at $SA[i+1]$ if and only if the k-mer is a common prefix of two suffixes at $SA[i]$ and $SA[i+1]$, i.e.,

$$k \leq lcp(i, i+1).$$

Repeat checking the above inequality until it no longer holds. If we confirm that $k \leq lcp(i+j, i+1+j)$ holds for $j = 0, \ldots, l$, the concurrence frequency

```
public static int[] occurrence_frequency(int[] SA, int[] LCP, int k){
    int[] freq = new int[LCP.length];
    int runLength = 1;
    for(int i = 0; i < LCP.length; i++){  // Checks if k > lcp(i,i+1) or not.
        if((i < LCP.length-1 && k > LCP[i+1]) || i == LCP.length-1){
            // The suffix at SA[i+1] is new, or the end of LCP is hit.
            // Store the occurrence frequency of the current k-mer.
            for(int j = i+1-runLength; j <= i; j++)
                freq[SA[j]] = runLength;
            runLength = 1;  // Initialize for the new k-mer.
        }else // Continue to search the current k-mer.
            runLength++;
    }
    return freq;
}
```

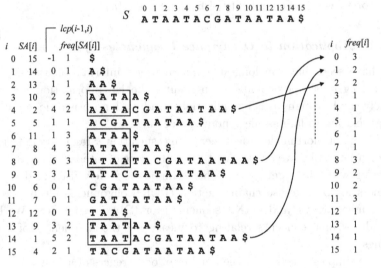

Fig. 4.10 The upper part presents a program for computing concurrence frequencies of all k-mers in the target string. The lower section illustrates its operation when computing occurrence frequencies of 4-mers for the current example.

of the k-mer is $l + 2$. Otherwise if $k > lcp(i, i+1)$, the k-mer appears only once, and the k-mer in the next position $SA[i + 1]$ must appear first in the suffix array, which allows the program to initiate calculation of the occurrence frequency of the next k-mer. Figure 4.10 presents a program that implements the above method on the current example. The program puts the occurrence frequency of the k-mer at $SA[i]$ into $freq[SA[i]]$. Note that it is sufficient to scan the LCP table only once to finalize the computation.

4.5.2 Application to the Longest Common Factors

The occurrence frequency is a basic measure for evaluating the uniqueness of k-mers. However, a k-mer string of occurrence frequency one could still be highly similar to another string if the two strings share a long stretch of matching characters called a *common factor*.

Definition 4.5 Let S be the target string, and Q be a substring of S. A *common factor* is a string shared between Q and any other substring of S that does not overlap with Q. Let $lcf(S,Q)$ denote the length of the longest common factor.

If we allow a substring of S that overlaps with Q, comparing Q with itself provides the longest common factor trivially. Even if we exclude Q itself from consideration, shifting Q by one position yields a substring that shares a long common factor with Q. We also assume that Q occurs only once in S; otherwise, it is obvious that $lcf(S,Q)$ is the length of Q when Q is compared to its copy.

For example, let $S = $ ATAATACGATAATAA\$ and $Q = $ ATAATAC. The longest common factor ATAATA between S and Q also occurs in ATAATAA\$, and it covers the first six characters of ATAATAC. Since the length of the longest common factor indicates the possibility of cross-reaction with off-target strings, it is one of standard criteria for checking the uniqueness of k-mers.

Suppose that Q ranges from l to r in S. Observe that the longest common factor of Q and S must be the longest common prefix between a suffix of Q and a suffix of S. If the suffix of Q starts at position k in S, it is sufficient to consider the longest common prefix between the suffix of Q at $k(= SA[ISA[k]])$ and its neighboring suffix at $SA[ISA[k] - 1]$ or $SA[ISA[k] + 1]$.

Note also that the longest common factor must terminate at the end of Q, and hence its length must be less than or equal to $r-k+1$. Consequently, $lcf(S,Q)$ maximizes

$$\min(r - k + 1, \ \max(lcp(ISA[k] - 1, ISA[k]), lcp(ISA[k], ISA[k] + 1)))$$

for $k = l, \ldots, r$. $lcf(S,Q)$ is computable in time proportional to the length of Q. It is straightforward to implement the above idea, but care has to be taken that $ISA[k] + 1$ does not exceed the boundary of SA.

4.6 Suffix Array Construction – Doubling

Sorting suffixes of a long string could be a computationally intensive task, if some suffixes have a long prefix in common. However, most suffixes can be ordered by comparing their short prefixes. Thus, it is reasonable to start examining the first symbol of each suffix and to extend the prefix to be checked. This approach allows us to generate an approximate suffix array in which suffixes are sorted according to prefixes of length h or less.

Definition 4.6 Let S be an input string such that each element except the last is a positive integer and the last element is zero, and let h be a positive integer. The *h-ordering* of suffixes is the list of suffixes that are sorted according to their prefixes of length h. An *approximate suffix array* SA_h of S is an array such that $SA_h[i] = k$ if and only if the i-th suffix in the h-ordering starts at position k in S.

Figure 4.11 presents examples of approximate suffix arrays. Symbols in gray boxes are not stored in approximate suffix arrays but are displayed for ease of understanding the behavior of the doubling technique. SA_1 is obtained from the 1-ordering of suffixes in which suffixes are sorted by the first element of each suffix. For example, nine suffixes have A as the first element. In addition, for the ease of understanding, they are ordered according to their starting positions in the input string, which is not required in the definition of the h-ordering. We must check the subsequent elements to order these suffixes lexicographically. However, suffixes starting at positions 6 and 7 have their unique first elements C and G. We do not need to compare these two suffixes to others, and we are able to determine their ranks at this point.

To settle the lexicographic order of suffixes starting with A and T, the 2-ordering is calculated, and the result is given in SA_2. Since prefixes of length 2 are compared, the block of suffixes with A are partitioned into four subparts that have respective prefixes A$, AA, AC, and AT, while the block of suffixes with T is still undivided in SA_2. In the next step, one may attempt to compute the 3-ordering, but we can proceed to consider the 4-ordering because in a prefix of length 4, e.g., $b_0 b_1 b_2 b_3$, ranks of both $b_0 b_1$ and $b_2 b_3$ can be calculated from the inverse suffix array of the 2-ordering, which will be explained shortly, and the combination of these two ranks gives the 4-ordering of suffixes.

This idea is called the *doubling technique* of Karp, Miller, and Rosenberg [43], which Manber and Myers utilized to efficiently construct suffix arrays

Suffix Arrays

0	1	2	3	4	5	6	7	8	9	10	11	12	13	14	15
A	T	A	A	T	A	C	G	A	T	A	A	T	A	A	$

SA_1	$S[SA_1[i]]$	SA_2	$S[SA_2[i], SA_2[i]+1]$	SA_4	$S[SA_4[i], SA_4[i]+3]$	SA_8	$S[SA_8[i], SA_8[i]+7]$	SA	
0	15 $	0	15	0	15	0	15	0	15
1	0 A	1	14 A $	1	14	1	14	1	14
2	2 A	2	2 A A	2	13 A A $	2	13	2	13
3	3 A	3	10 A A	3	2 A A T A	3	10 A A T A A $	3	10
4	5 A	4	13 A A	4	10 A A T A	4	2 A A T A C G A T	4	2
5	8 A	5	5 A C	5	5	5	5	5	5
6	10 A	6	0 A T	6	0 A T A A	6	11 A T A A $	6	11
7	11 A	7	3 A T	7	8 A T A A	7	8 A T A A T A A $	7	8
8	13 A	8	8 A T	8	11 A T A A	8	0 A T A A T A C G	8	0
9	14 A	9	11 A T	9	3 A T A C	9	3	9	3
10	6 C	10	6	10	6	10	6	10	6
11	7 G	11	7	11	7	11	7	11	7
12	1 T	12	1 T A	12	12 T A A $	12	12	12	12
13	4 T	13	4 T A	13	1 T A A T	13	9 T A A T A A $	13	9
14	9 T	14	9 T A	14	9 T A A T	14	1 T A A T A C G A	14	1
15	12 T	15	12 T A	15	4 T A C G	15	4	15	4

Fig. 4.11 Doubling technique.

[59, 60]. In general, if the h-ordering is available, sorting prefixes of length $2h$ in suffixes can be performed by using the former half of each prefix as the primary key and the latter half as the secondary key. In Figure 4.11, the doubling technique is used to build SA_8 from SA_4. Observe that all suffixes are finally ordered lexicographically in SA_8. In this way, the sequence of SA_1, SA_2, SA_4, and SA_8 approximates to $SA = SA_8$.

In the worst case, we must check all prefixes of suffixes; consider the example when all elements in an input string are the same, e.g., A. If the input string is of length n, to compute the suffix array we need to generate $SA_{2^0}, SA_{2^1}, \ldots, SA_{2^k}$ such that 2^k minimizes $n < 2^k$. As $k = \lceil \log_2 n \rceil$, the number of approximate suffix arrays built during the computation is at most $1 + \lceil \log_2 n \rceil$. Give such an example that the number becomes $1 + \lceil \log_2 n \rceil$. It remains to implement the procedure of building $SA_{2^{i+1}}$ from SA_{2^i}. In what follows, we present the Larsson-Sadakane algorithm [54] that outputs suffix arrays in $O(n \log_2 n)$ time.

4.7 Larsson-Sadakane Algorithm

The initial step of the doubling technique is to sort the first letters of all suffixes. To this end, the radix sort algorithm should be used if the number of distinct first letters is smaller than the length of the input string. Otherwise, the randomized quick sort is a method of choice.

To implement the doubling technique, given the h-ordering of prefixes of all suffixes, we attempt to determine the ordering of suffixes having the same prefix of length h in common by looking at succeeding substrings of length h that immediately follow the prefix. All these substrings are prefixes of some suffixes, and hence, their ordering may be decided according to the h-ordering. However, prefixes in the h-ordering are not fully ordered; some are equally ranked.

Definition 4.7 Suffixes that have the same prefix of length h and are therefore ranked equally in the h-ordering are noted as having the same h-rank and are in the same h-group. The starting positions of suffixes in the same h-group are stored consecutively in SA_h; suppose that the indexes of the consecutive block ranges from l to r in SA_h. The h-rank of the h-group is defined as the last index r. An *approximate inverse suffix array* ISA_h is an array such that $ISA_h[SA_h[i]] = r$ for $i = l, \ldots, r$.

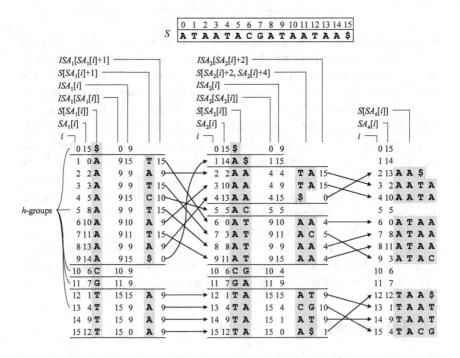

Fig. 4.12 Operation of the Larsson-Sadakane algorithm.

```
public static int[] suffixArray_LS(int[] S){
    int n = S.length;  int[] SA = new int[n];  int[] ISA = new int[n];
    // Basic step.
    radixSort_LS(S, SA, ISA);
    // Iterative step.
    boolean iteration; int h=1;
    do{ iteration = false;
        for(int i=0; i<n; i = ISA[SA[i]]+1)
            if(i < ISA[SA[i]]){  // If some h-groups are divided.
                ternary_split_quicksort_LS(SA, i, ISA[SA[i]], ISA, h);
                iteration = true;   // Repeat the while-loop.
            }
        h = h*2;
    }while(iteration);
    return SA;
}
public static void radixSort_LS(int[] S, int[] SA, int[] ISA){
    // Treat each element as one digit, execute radix sort, and update ISA.
    int n = S.length;   int m=0;
    for(int i=0; i<n; i++)
        if(m < S[i]) m = S[i]; // Calculate the maximum.
    m++; // Total number of elements
    int[] count = new int[m];
    for(int i=0; i<m; i++){ count[i]=0; }
    for(int i = 0; i < n; i++){ count[S[i]]++; }
    for(int i=1; i<m; i++){ count[i] += count[i-1]; }
    for(int i=n-1; 0<=i; i--){ ISA[i] = count[S[i]]-1; }  // update ISA
    for(int i = n-1; 0 <= i; i--){ SA[count[S[i]]-1] = i; count[S[i]]--; }
}
```

Fig. 4.13 The upper portion shows the top-level of the Larsson-Sadakane algorithm, while the lower part presents the subroutine for the basic step.

Figure 4.12 illustrates h-groups partitioned by horizontal lines. ISA_h is built by setting $ISA_h[SA_h[i]] = r$ for $i = l, \ldots, r$ where l and r are boundaries of the range of each h-group in the above definition. We are ready to sort substrings of length h that immediately follow the common prefix in an h-group because their ordering can be determined by the order described in $ISA_h[SA_h[i] + h]$. For sorting elements in $ISA_h[SA_h[i] + h]$, the Larsson-Sadakane algorithm utilizes the ternary split quick sort that separates all pivot values as a block from the others, making it possible to identify one $2h$-group without its being divided. The reason why the randomized quick sort is not chosen is that it may fail to categorize all the pivot values into one group, making it inappropriate for use in this application.

Figures 4.13 and 4.14 present the Larsson-Sadakane algorithm. The

```
public static void ternary_split_quicksort_LS
                (int[] SA, int l, int r, int[] ISA, int h) {
    if(l > r) return;  // Exit when the empty range is input.
    if(l == r){ // Set ISA[SA[l]] to l when the input is a singleton.
        ISA[SA[l]] = l; return;
    }
    // Choose a pivot from ISA[SA[]+h].
    int v = ISA[SA[l+(int)Math.floor(Math.random()*(r-l+1))]+h];
    int i = l; int mi = l; int j = r; int mj = r; int tmp;
    // Compare values according to ISA[SA[]+h].
    for (;;) {
        for(; i <= j && ISA[SA[i]+h] <= v; i++)
            if(ISA[SA[i]+h] == v){
                tmp = SA[i]; SA[i] = SA[mi]; SA[mi] = tmp; mi++; }
        for(; i <= j && v <= ISA[SA[j]+h]; j--)
            if(ISA[SA[j]+h] == v){
                tmp = SA[j]; SA[j] = SA[mj]; SA[mj] = tmp; mj--; }
        if (i > j)   break;
        tmp = SA[i];  SA[i] = SA[j];  SA[j] = tmp; i++; j--;
    }
    for(mi--, i--; l <= mi; mi--, i--){
        tmp = SA[i];   SA[i] = SA[mi];   SA[mi] = tmp; }
    for(mj++, j++; mj <= r; mj++, j++){
        tmp = SA[j];   SA[j] = SA[mj];   SA[mj] = tmp; }
    // Recursive call.
    ternary_split_quicksort_LS(SA, l, i, ISA, h);
    for(int k = i+1; k < j; k++) ISA[SA[k]] = j-1;  // update ISA
    ternary_split_quicksort_LS(SA, j, r, ISA, h);
}
```

Fig. 4.14 The Larsson-Sadakane algorithm modifies the ternary split quick sort to incorporate and manipulate SA and ISA information. The differences are annotated with comments.

upper half in Figure 4.13 shows the top level of the program that initially sorts first elements of suffixes while computing array ISA by executing the radix sort subroutine in the lower half of the figure. Subsequently, the program identifies h-groups with more than one suffix of the same h-rank by scanning ISA[SA[i]] to detect index i such that i is strictly less than the rightmost index of its h-group ISA[SA[i]]. The program then attempts to sort elements of ISA[SA[i]+h] in individual h-groups by calling the ternary split quick sort presented in Figure 4.14 that updates ISA during the computation.

Performance acceleration of the algorithm can be achieved if the number of symbols in the input string is small. Typical examples are nucleotide strings. For example, transforming all 4-mer nucleotide substrings into 1-

byte integers allows us to radix sort 4-mer substrings at once to yield an approximate suffix array SA_4 without building SA_1 and SA_2. In general, compressing h-mer substrings makes the direct construction of SA_h possible, and the Larsson-Sadakane algorithm incorporates the idea.

The Larsson-Sadakane algorithm needs $4n$ bytes to store SA and additional $4n$ bytes for ISA if integers are represented in 4 bytes. Next, we analyze the worst-case time complexity of the algorithm. Partitioning h-groups dominates the overall performance. Although Figure 4.12 illustrates how h-groups are calculated for $h = 1, 2, 4$, it does not depict the precise steps of partitioning, making it difficult to analyze the worst-case time complexity. Figure 4.15 shows the tree in which nodes represent sets of h-long prefixes and directed arcs indicate the initial division by the radix sort and subsequent ternary splits. Each node is labeled with the pair of the length of prefixes and the number of prefixes in the set. Here, we assume that the true median is selected as the pivot element in the Larsson-Sadakane algorithm. We can fulfill this requirement in time linear to the number of elements in the set, using the algorithm of Schönhage, Paterson, and

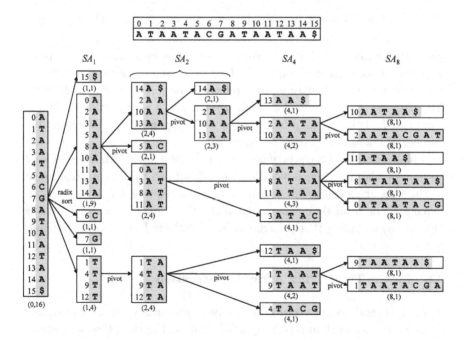

Fig. 4.15 Precise steps of partitioning in the Larsson-Sadakane algorithm.

Pippenger [88] in place of random selection, though the linear-time algorithm hardly improves the performance of the algorithm due to increased constant factors.

The ternary split sort generates two types of child node: nodes containing the pivot values, called *pivot nodes* and others called *non-pivot nodes*. In Figure 4.15, pivot nodes have incoming arcs labeled with "pivot." Selecting the true median as the pivot guarantees that the size of any non-pivot node is less than half the size of its parent node. On the other hand, any pivot node has the set of the identical h-long prefixes, which forms one h-group. Therefore, if the h-group has more than one prefix, it is immediately divided into subsets of $2h$-long prefix in the child nodes of the parent pivot node. Observe that the non-pivot and pivot nodes in Figure 4.15 have these respective properties.

Any path from the root to a leaf has the root and one node that the radix sort generates, and it may also have some pivot and non-pivot nodes. For example, in Figure 4.15, consider the path from the root to the leaf "10 AATAA$" that has the following chain of labels:

$$(0,16) \to (1,9) \to (2,4) \to (2,3) \to (4,2) \to (8,1)$$

Any path has at most $\lceil \log_2 n \rceil$ pivot nodes because child nodes of any pivot node with h-long prefixes contain $2h$-long prefixes, and hence the path has at most one pivot node with h-long prefixes for each $h = 2^1, 2^2, \ldots, 2^{\lceil \log_2 n \rceil}$. Any path also has at most $\lceil \log_2 n \rceil$ non-pivot nodes because the size of any non-pivot node is less than half the size of its parent node. Therefore, the number of nodes in any path is $O(\log_2 n)$.

Finally, note that the sets of prefixes in nodes of any depth from the root are non-overlapping. Since partitioning each set can be done in time linear to the size, partitioning all sets of any depth requires $O(n)$ time. Consequently, the total time complexity of the Larsson-Sadakane algorithm is $O(n \log_2 n)$ assuming that the true median is selected as the pivot element.

4.8 Linear-Time Suffix Array Construction

In 2003, several groups reported recursive algorithms for constructing the suffix array that work in time proportional to the length of the input string [42, 50]. These algorithms adopt the idea of the divide-and-conquer approach:

(1) Divide suffixes into two classes, e.g., suffixes starting at positions $i \bmod 3 \neq 0$ and the others.
(2) Construct the suffix array of the first class by recursive call.
(3) Use the result to generate the suffix array of the second class.
(4) Merge the two suffix arrays into one, and return the result.

Among several proposals, we here introduce the linear-time algorithm designed by Kärkkäinen and Sanders [42]. Given an input string of length n, this algorithm builds the suffix array of $2n/3$ suffixes in Step 2. The crux of the algorithm is that it performs Steps 3 and 4 in time cn for a constant c. Let $T(n)$ denote the computation time of the overall execution. The recursive call in Step 2 takes $T(2n/3)$, and hence we have the recurrence $T(n) = T(2n/3) + cn$. Solving this gives $T(n) \approx 3cn$, and therefore the time complexity of the algorithm is $O(n)$.

Figure 4.16 shows the operation of Kärkkäinen-Sanders algorithm on the input string ATAATACGATAATAA put into variable S. The algorithm assumes that all elements in the input string are positive integers. Note that we do not append the terminal symbol $ explicitly at the end of the input string, but the algorithm instead automatically pads the input with three zeros, which are strictly less than any element in the input string. Appending three zeros avoids array boundary checking, thereby reducing the number of conditional statements executed in the program.

To construct the suffix array named SA12 of suffixes that start at position i mode $3 \neq 0$, we store their starting positions into array S12.

$$S12[j] = i = f(j) = \begin{cases} 3j + 1 & \text{if } j < n0 = (n+2)/3 \\ 3(j - n0) + 2 & \text{otherwise} \end{cases}$$

S12[j] represents the starting position i=f(j) for the suffix in the input S. The first half stores such indexes that $i \bmod 3 = 1$, e.g., 1, 4, 7, ..., and the latter half contains $i \bmod 3 = 2$, e.g., 2, 5, 8, ... We then radix sort first triplets of individual suffixes so that all the triplets are ranked and put into rank12. During this sorting, first triplets are not actually put into S12 but are processed by accessing the input S. This radix sort requires a counter of a size that is equal to the number of nonnegative integers including zero in S. The size of the counter is therefore at most the length of S. For example, if the input consists of four nucleotide letters, the counter size equals four plus one.

We then replace starting positions of suffixes in S12 with ranks of the first triplets of suffixes. For example, the first triplet TAA in the suffix

76 Large-Scale Genome Sequence Processing

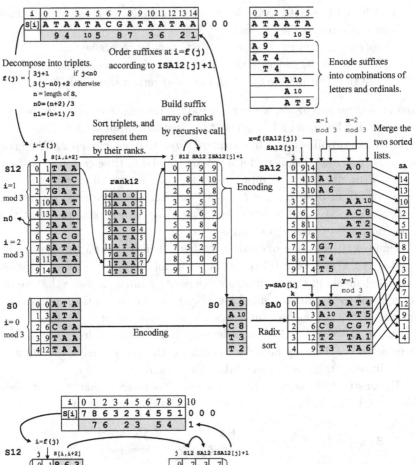

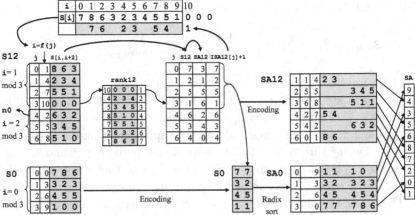

Fig. 4.16 Operation of Kärkkäinen-Sanders algorithm. The gray elements are not stored in arrays but are calculated during the computation. The upper and lower illustrations present the processes of building suffix arrays of ATAATACGATAATAA and 7863234551, respectively.

starting at position 1 of S is ranked 7 among all the triplets, and hence 7 is assigned to S12[0]. This replacement transforms S12 into the list of ranks of triplets, 7863234551. It is important to observe that the suffix array of string 7863234551 gives the suffix array of suffixes in the original S12.

If all triplets are ranked differently, it is almost straightforward to order suffixes in S12 according to the ranks of triplets; however, more than one triplet can have the same rank. For example, AAT occurs at suffixes of position 2 and 10. In this case, we call the algorithm recursively to generate the suffix array 9435687201 for string 7863234551. Since the length of 7863234551 is 10, which is the two-thirds the length of the input S, the original problem is partitioned into a smaller problem.

Next we consider how to build the suffix array of suffixes that start at position i mod $3 = 0$. First, the starting positions of these suffixes are put into S0. We then treat each suffix as the pair of its first element and the rank of the suffix next to the first element according to the order defined by SA12. For instance, the suffix starting at position 0 is represented by A9, because its first element is A and its next suffix is ranked 9. This encoding allows us to radix sort pairs for suffixes in S0 to build the suffix array SA0 for S0, which takes time proportional to the size of S0.

Having the two sorted lists of suffixes in SA12 and SA0, the final step is to merge them. Care must be taken to compare the two elements in these two lists. A suffix in SA12 that starts at x can be represented in one of the following two forms:

- If x mod $3 = 1$, the suffix starting at x+1 is ranked, making it possible to represent the suffix as the pair of the first element and the rank of the following suffix, e.g., A1 for x=13.
- If x mod $3 = 2$, the suffix starting at x+1 is not ranked because x+1 mod $3 = 0$. We therefore denote the suffix by the triplet of the first and second elements and the rank of the following suffix, e.g., AC8 for x=5.

In the former case, a suffix cannot be represented by the triplet because the rank of the suffix starting from x+2 is missing; e.g., consider the case when x=1. Similarly, in the latter case, we cannot express a suffix by a pair.

However, a suffix in SA0 can be denoted by both representations. For example, the suffix starting at 0 can be expressed by either A9 or AT4, making it possible to compare it to any suffix in SA12. Merging the two sorted lists yields the suffix array of the original input string in S. This step is straightforward and takes time proportional to the sum of the lengths of the two lists.

We here note a special treatment to process strings, e.g., 7863234551, which end at positions that are multiples of three. The lower half of Figure 4.16 illustrates how the recursive call of the algorithm processes 7863234551 to build its suffix array. All elements in S0 are compared to those in SA12, and during this step, the rank of the suffix next to each element of S0 must be examined. However, the last element in S0, position 9, does not have its next element in the input string. To resolve this issue, the program considers the virtual suffix 000 starting at position 10, and it adds 000 to S12. Since 000 must be smaller than all suffixes in the input string, this method ensures that the suffix next to the last element of the input string is the smallest. This particular treatment is necessary only if the position of the last element is a multiple of three, and n0 is greater than n1 by one. Otherwise, n0 = n1, and it is not necessary to add 000 to S12.

The program that implements Kärkkäinen-Sanders linear-time suffix array construction algorithm is displayed in Figures 4.17 and 4.18. The first half of Figure 4.17 presents a straightforward implementation of the method discussed so far. However, the latter half of Figure 4.18 needs some explanation. First, to sort elements ISA12[j]+1 that are displayed in the second column (the least significant digit) of S0, the program utilizes the property that the order of these elements is available in SA12. Second, during the step of merging two sorted lists SA12 and SA0, it is elaborate to access to the rank of the suffix that starts at the last position, e.g., x+1 and x+2, of a pair or a triplet. Indexes for accessing ISA12 have to be set properly depending on whether x is 1 mod 3 or 2 mod 3.

The program in Figures 4.17 and 4.18 is not space efficient. More precisely, S12, rank12, S0, and SA0 are integer arrays of lengths $2n/3$, $2n/3$, $n/3$, and $n/3$ respectively. The size of the count array is dependent of the number of symbols to use but could be n in the worst case when all input symbols are distinct. If the input string consists of four nucleotides, the count array has four integers in the initial call of the procedure. However, the number of symbols tends to increase as the procedure is called recursively, while the size of the input string becomes exponentially small rapidly. In the worst case, the initial call demands integer arrays of $3n$ in length, and subsequent recursive calls need integer arrays of length $6n$ $(= 3n \times (2/3 + (2/3)^2 + (2/3)^3 + \ldots))$. Therefore, the program may use $36n$ $(= 4 \times (3n + 6n))$ bytes in the worst case if integers are represented by 4 bytes.

```
public static void suffixArray_KS(int[] S, int n, int[] SA, int numRanks){
   // Elements in S must be positive integers except the last three zeros.
   // n = S.length - 3.0 < n.numRanks = number of symbols in S.
   int n0=(n+2)/3; int n1=(n+1)/3; int n2=n/3; int n02 = n0+n2;
   // If n0=n1+1, an extra element of S12 at position n0-1 denotes
   // the triplet "000" of S at position n+1.
   int[] S12 = new int[n02+3]; S12[n02]=S12[n02+1]=S12[n02+2]=0;
   for(int j=0; j < n02; j++) S12[j] = f(j, n0);
   int[] rank12 = new int[n02];
   int[] count  = new int[numRanks];
   // Sort triplets by radix sort.
   for(int d=2; 0<=d; d--){
       for(int i=0; i < numRanks; i++){ count[i]=0; }
       for(int j=0; j < n02; j++) count[S[S12[j]+d]]++;
       for(int i=1; i < numRanks; i++) count[i] += count[i-1];
       for(int j=n02-1; 0 <= j; j--)
           rank12[--count[S[S12[j]+d]]] = S12[j];
       for(int j=0; j < n02; j++) S12[j] = rank12[j];
   }
   // Represent triplets by their ranks.
   int rank = 1;
   for(int j=0; j < n02; j++){
       if(0<j){ // Increment rank if neighboring triplets are not equal.
           int x = rank12[j-1]; int y = rank12[j];
           if(compare(S[x],S[x+1],S[x+2],S[y],S[y+1],S[y+2])!=0) rank++;
       }        // rank12[j] indicates index i in S.
       if(rank12[j]%3==1) S12[rank12[j]/3]    = rank;   // i = 1 mod 3
       else               S12[rank12[j]/3+n0] = rank;   // i = 2 mod 3
   }
   // Generate the suffix array SA12 and its inverse ISA12 for S12.
   int[] SA12 = rank12;  // Reuse rank12 as it is not needed at this point.
   if(rank == n02)       // If no duplicates, perform radix sort.
       for(int j=0; j<n02; j++) SA12[S12[j]-1]=j;
   else                  // Otherwise, recursive call.
       suffixArray_KS(S12, n02, SA12, rank+1);
   int[] ISA12 = S12;    // Reuse S12 array as it is not needed.
   for(int j=0; j<n02; j++) ISA12[SA12[j]] = j+1;  // Increment ISA12[j].
```

Fig. 4.17 First half of the Kärkkäinen-Sanders linear-time suffix array construction algorithm.

One can reduce the space used by the program in a couple of ways. The algorithm merges two arrays SA12 and SA0 and puts the result into SA in the final step. Before the end, the space in SA that occupies $4n$ bytes can be used as working space. Similarly, some other arrays are not used simultaneously and can be shared by statements in the program [79].

```
// Sort elements in the second column of S0.
int[] S0 = new int[n0];
for(int j=0, k=0; j < n02; j++)  // ISA12[j]+1 are sorted in SA12.
    if(SA12[j]<n0) S0[k++] = 3*SA12[j]; // i=f(j) and i=1 mod 3.
// Sort elements in the first column of S0 by radix sort.
int[] SA0 = new int[n0];        // Sorted list of indexes in S0
for(int i=0; i < numRanks; i++) count[i]=0;
for(int k=0; k < n0; k++)       count[ S[S0[k]] ]++;
for(int i=1; i < numRanks; i++) count[i] += count[i-1];
for(int k=n0-1; 0 <= k; k--) SA0[--count[S[S0[k]]]] = S0[k];
// Merge two sorted lists SA12 and SA0 to generate SA.
int j, k, l;
for(j=n0-n1, k=0, l=0; j<n02 && k<n0; l++){
    int comparison; int x = f(SA12[j],n0); int y = SA0[k];
    if(SA12[j] < n0){ // x = 1 mod 3
        comparison = compare(S[x], ISA12[SA12[j]+n0],
                             S[y], ISA12[y/3]);
        // If y is the last position of S, y/3=n0-1 and n0=n1+1.
        // S12[y/3] denotes "000", and ISA[y/3] must be 0.
    }else{ // x = 2 mod 3
        comparison = compare(S[x], S[x+1], ISA12[SA12[j]-n0+1],
                             S[y], S[y+1], ISA12[y/3+n0]);
        // If x is the last position of S, S[x+1] must be 0.
        // The former triplet must be lower than the latter.
    }
    // S[x]... < S[y]... if and only if comparison == -1
    if(comparison == -1){ SA[l]=x; j++; }
    else{ SA[l]=y; k++; }
}
for(; j < n02; l++, j++) SA[l] = f(SA12[j],n0);
for(; k < n0;  l++, k++) SA[l] = SA0[k];
}
public static int f(int j, int n0){
    if(j < n0) return 3*j+1; else return 3*(j-n0)+2;
}
public static int compare(int i0, int i1, int j0, int j1){
    // return -1, 0, 1 if i < j, i == j, and i > j, respectively.
    if(i0 < j0) return -1;
    else if(i0 == j0){
        if(i1 < j1) return -1; else if(i1 == j1) return 0; else return 1; }
    else return 1;  // i0 > j0
}
public static int compare(int i0, int i1, int i2, int j0, int j1, int j2){
    // return -1, 0, 1 if i < j, i == j, and i > j, respectively.
    if(i0 < j0)        return -1;
    else if(i0 == j0) return compare(i1, i2, j1, j2);
    else              return 1;
}
```

Fig. 4.18 Latter half of the Kärkkäinen-Sanders algorithm and its subroutines.

4.9 A Note on Practical Performance

In this chapter, we show that the Larsson-Sadakane algorithm computes the suffix array in $O(n \log_2 n)$ time and requires $8n$ bytes, while the Kärkkäinen-Sanders linear-time algorithm needs more space. Furthermore, many other algorithms have been developed for suffix array construction. In particular, lightweight algorithms are promising because they need less space, e.g., $5n$ or $6n$ bytes, though they run in $O(n^2 \log_2 n)$ time [15, 61, 92]. It appears that there is a trade-off between time and space when using these algorithms.

However, we have seen that, in comparing various sorting algorithms, the worst-case time complexity does not always provide an accurate measure for assessing the practical performance of a sorting algorithm. Therefore, it is worthwhile to compare the practical performance of suffix array construction algorithms. To this end, Puglisi, Smyth, and Turpin conducted extensive tests [80]. According to their experiments, the Manzini-Ferragina algorithm, an $O(n^2 \log_2 n)$ time lightweight algorithm, used the least memory ($5n$ bytes) and outperformed other algorithms on average. The Larsson-Sadakane algorithm was slightly slower than the Manzini-Ferragina algorithm, but was faster than the Kärkkäinen-Sanders algorithm on average.

Problems

Problem 4.1 Implement a program that searches the suffix array of the target string for the rightmost position of the block of suffixes that has the given query as their prefixes with or without using the longest common prefix information.

Problem 4.2 Design an algorithm for finding the longest common substring of two strings S and S' in $O(|S| + |S'|)$ time.

Problem 4.3 If a suffix of S is identical to a prefix of S', the suffix is called a *suffix-prefix* of S and S'. Design an algorithm for finding the longest suffix-prefix of S and S' in $O(|S| + |S'|)$ time.

Problem 4.4 Prove Proposition 4.1; i.e., for $i + 1 < j$,

$$lcp(i, j) = \min\{lcp(k - 1, k) \mid k = i + 1, \ldots, j\}.$$

Problem 4.5 Present such a target string of length n that the number of approximate suffix arrays built during the doubling technique computation becomes $1 + \lceil \log_2 n \rceil$.

Problem 4.6   The Larsson-Sadakane algorithm incorporates the idea of compressing h-mer substrings into 1-byte or one-word integers to construct SA_h directly. Design and implement such a program.

Chapter 5

Space-Efficient String Search

Lookup tables and suffix arrays are generated by preprocessing the target sequence in order to accelerate the performance of answering queries. If the target sequence is very long, these data structures are extremely space intensive and are likely to get too large to fit into the main memory. For example, if one preprocesses the human genomic sequence of about 3×10^9 nucleotides and generates a direct-address table of 15-mer substrings, about 16 gigabytes of main memory is required.

While the target sequence, such as genomic sequences, could be very long, typical query sequences are short when we look for exact matches. To study genome evolution, one may need to query a long genomic sequence against other genomes, approximate string matching rather than exact matching is essential; this issue will be discussed later in this book.

When queries are short, there are several algorithms that preprocess short queries rather than long target sequences in order to gain efficiency. While the generation of large lookup tables is computationally costly, preprocessing short queries is inexpensive, and hence, queries of various lengths are much easier to handle. Here we introduce three major techniques. All of these algorithms were developed in the 1970s when main memory was so precious that only several megabytes were available, even on huge mainframe computers that occupied a large floor area.

5.1 Rabin-Karp Algorithm

Figure 1.3 presents the simple but naive string search algorithm that transforms strings into integers and compares the k-mer integer of the query with k-mer integers of all k-mer substrings in the target. One problem with this approach is that a k-mer integer could be very large; e.g., a 100-mer integer may amount to $4^{100} = 2^{200}$, which cannot be coded in 32 or 64 bits.

Encoding substrings into 200 bits in this case is likely to be superfluous. For instance, given the human genome sequence of length 3×10^9, the number of 100-mers is about 3×10^9, which is much smaller than $2^{200} \approx 10^{60}$. In such cases, mapping 100-mer integers in space of a higher dimension into integers in space of a lower dimension may avoid heavy arithmetic operations. For this purpose, it is useful to employ a hash function of the form hash$(x) = x$ mod p, where p is a prime number and is the lower dimension. For example, you can use the Mersenne prime number $2^n - 1$ for $n = 17, 19, 31$.

Unfortunately, two major problems make it difficult to incorporate this idea into the brute-force algorithm in Figure 1.3. First, recall that, the l-mer integer of $T[k+1, k+l]$ is calculated from the l-mer integer of $T[k, k+l-1]$ using the following formula:

$$encode(T[k+1, k+l]) \\ = (encode(T[k, k+l-1]) - encode(T[k]) \cdot 4^{l-1}) \cdot 4 + encode(T[k+l]).$$

One can compute $encode(T[k+1, k+l])$ and then take mod p, which is likely to yield huge integers that are much greater than p during the computation. To avoid this, the following equation allows us to take mod p in earlier steps, which avoids generating huge integers and computes the hash value of the next l-mer integer of $T[k+1, k+l]$ from that of previous string $T[k, k+l-1]$.

$$encode(T[k+1, k+l]) \bmod p \\ = (((encode(T[k, k+l-1]) \bmod p) - encode(T[k]) \cdot (4^{l-1} \bmod p)) \cdot 4 \\ + encode(T[k+l])) \bmod p.$$

The second issue is that while hash$(x_1) \neq$ hash(x_2) ensures $x_1 \neq x_2$, the converse is not true. We must cope with this collision problem, i.e., that a hash function may associate more than one distinct number with the same hash value. Assuming that x maps to a random number ranging from 0 to $p - 1$, the probability that two distinct numbers have the same hash value is $1/p$, and hence, the collision problem rarely occurs, although we must anticipate this situation. When the query and a substring of the same length have identical hash values, we need to determine if the two strings are equal by comparing all pairs of two corresponding letters. If the query string occurs many times in the target string, this extra confirming step would be executed to ensure that most of the comparisons are successful.

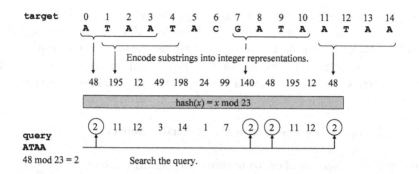

```
public static void rkSearch( int[] target, int[] query ) {
    int targetLen = target.length;
    int queryLen  = query.length;
    int primeNumber = 524287;  // 2^{19}-1, a Mersenne prime number.
    int intQuery = 0;          // The hash value of the query.
    int tmp = 0;               // The hash value of a temporary substring.
    int c   = 1;               // The power of 4 to queryLen mod primeNumber.
    for(int i = 0; i <= (targetLen - queryLen); i++){
        if(i == 0){  // Initialization.
            for(int j = 0; j < queryLen; j++){
                intQuery = (intQuery*4 + query[j]) % primeNumber;
                tmp = (tmp*4 + target[j]) % primeNumber;
                c = c*4 % primeNumber;
            }
        }else         // Iterative step.
            tmp = (tmp*4  - target[i-1]*c + target[i+queryLen-1])
                    % primeNumber;
        if(intQuery == tmp){
            int j;
            for(j = 0; j < queryLen && target[i+j]==query[j]; j++){}
            if(j == queryLen) // Print positions of query occurrences.
                System.out.print(i+" ");
        }
    }
    System.out.println();
}
```

Fig. 5.1 The upper half presents how the Rabin-Karp algorithm in the lower half operates on target and query.

If the number of occurrences is very small, however, the extra cost may be negligible.

Figure 5.1 presents the Rabin-Karp algorithm [44] and describes how it processes the running example. To resolve the conflict problem that 140 and 48 have the same hash value 2 in the figure, all pairs of letters in GATA and ATAA are compared. The program is almost identical to the brute-force algorithm in Figure 1.3 except that values modulo p are computed and precise string comparisons are made even if two hash values are equal.

5.2 Accelerating the Brute-Force String Search

We have thus far seen how to improve the performance of the brute-force string search algorithm in Figure 1.2 by encoding substrings into k-mer integers and by utilizing some additional data structures to accelerate access to occurrences of the query string. Instead of using k-mer integers, we attempt to improve the brute-force string search algorithm by eliminating unnecessary searches.

We will consider the following target and query strings in place of the running examples that we have used so far.

```
target   ATATATATATTAATATATT
query    ATATATT
```

This is because the brute-force search algorithm takes more unnecessary steps to process the above example as illustrated in the upper half of Figure 5.2. Two variables, i and j, scan the target and the query, respectively, and the leftmost two columns show ranges that individual variables take during the computation. Bold letters represent characters that are compared, and the gray boxes show disagreements between two letters in the same position of the target and the query. For example, the third row indicates that both i and j range from 0 to 6, and T in the gray box indicates its disagreement of A in the 6th position of the target string.

Many comparisons in the brute-force string search are redundant. We will show how to eliminate these redundancies to produce an ideal execution scenario in the lower half of Figure 5.2.

Let us examine the computation of the brute-force search in the upper half of the figure. After making comparisons in the 3rd row, we know that the 6-mer prefix of the query, ATATAT, occurs in the target. Thus, there is no way to compare target[1] and query[0] (= A) in the 4th row, which allows us to move to the computation in the 5th row immediately. The calculation in the 5th row can be also reduced, because we know that the 4-mer suffix of ATATAT and the 4-mer prefix of the query are identical, which

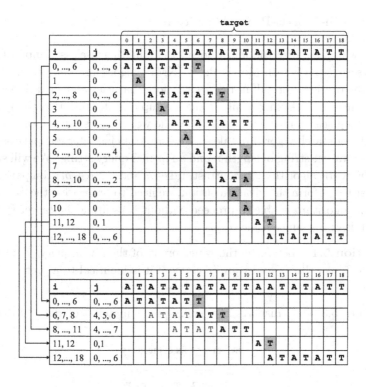

Fig. 5.2 The upper half illustrates how the brute-force string search algorithm in Figure 1.2 operates on the target string ATATATATATTAATATATT and query ATATATT. The lower half shows an ideal execution scenario after removing unnecessary and redundant comparisons in the upper half computation. In the lower half, only bold letters are compared as indicated by ranges of i and j.

makes it useless to compare them by consulting the target string and allows us to perform pairwise letter comparisons from where i = 6 and j = 4, as illustrated in the lower half of the figure. Similarly, the 6th row can be eliminated, and the 7th row can begin with positions i = 8 and j = 4, which is also shown in the 5th row of the ideal execution.

In the 7th row, all pairwise comparisons are successful, and the query ATATATT is confirmed to occur at the 4th position of the target. This information enables us to avoid all comparisons in the 8th to 13th rows without checking the target string. Observe that if the query matches the target from positions 4 to 10, a proper suffix of ATATATT must be identical to a prefix of the query ATATATT, which in this case does not hold.

5.3 Knuth-Morris-Pratt Algorithm

We will generalize the above idea of eliminating useless computations and describe the Knuth-Morris-Pratt (KMP) algorithm [49]. In the ideal execution of the lower half in Figure 5.2, note that the value of variable i is monotonically increasing, while it changes up and down during the brute-force string search in the upper half. In the worst-case scenario, the brute-force string search algorithm takes $m(n - m + 1)$ character comparisons to process a target string of length n and an m-mer query string. We will show that the KMP algorithm, in contrast, takes at most $2(n + m)$ comparisons. In order to perform the ideal execution, we need to preprocess the query so that the algorithm is able to proceed without consulting the target string. We will first introduce some terms.

Definition 5.1 Let P_i be the i-mer prefix of the given query. A proper suffix of P_i is called a *suffix-prefix match* if it is identical to a prefix of the query.

Figure 5.3, e.g., displays two suffix-prefix matches of query ATATAT.

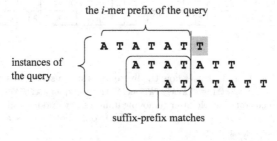

		0	1	2	3	4	5	6	7	8	9	10	11	12	13	14	15	16	17	18
i	j	A	T	A	T	A	T	A	T	A	T	T	A	A	T	A	T	A	T	T
0, ..., 6	0, ..., 6	A	T	A	T	A	T	**T**												
6, 7, 8	4, 5, 6					A	T	A	T	A	T	**T**								
8, 9, 10	2, 3, 4							A	T	A	T	**A**								
8, 9, 10	4, 5, 6					A	T	A	T	**A**	**T**	**T**		oversight						

Fig. 5.3 The upper figure shows that two suffixes of the 6-mer prefix in query ATATAT match the prefixes of the query. The longest suffix-prefix match indicates the length that the query should be shifted. The lower picture indicates that the failure to use the longest suffix-prefix match may lead to an oversight of query occurrences. For example, the query occurrence starting at the 4th position of the target, is overlooked.

Among them, the longest one should be selected for shifting the query because the failure to use the longest may overlook occurrences of the query in the target. The lower table in Figure 5.3 illustrates that using another shorter suffix-prefix match AT passes an occurrence of the query and starts comparison at a position where $i = 8$ and $j = 2$, which overlooks the occurrence of the query from the 4th position of the target.

Algorithm 5.1 Suppose that during the search of the target for the query, the i-mer prefix of the query matches the target, but the character followed by the i-mer disagrees with the corresponding letter in the target. The *KMP shift rule* moves the query by the length of the longest suffix-prefix match of the i-mer.

Figure 5.4 illustrates how the KMP shift rule operates. One might be concerned that the KMP shift rule overlooks query occurrences in the target, as illustrated in the figure. We can prove that the KMP shift rule enumerates all occurrences of the query. Otherwise there exists an unnoticed query occurrence in the target, as illustrated in Figure 5.4. This, however, would give another suffix-prefix match of the i-mer prefix that is longer than the longest suffix-prefix match used by the KMP shift rule, which is a contradiction.

In order to implement the KMP shift rule, we must calculate the length of the longest suffix-prefix match of each i-mer prefix in the query before-

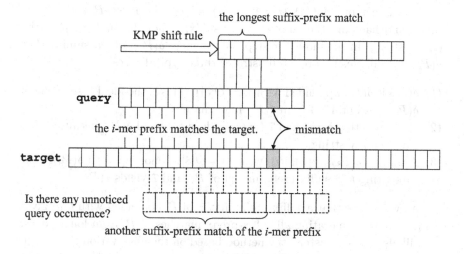

Fig. 5.4 The KMP shift rule.

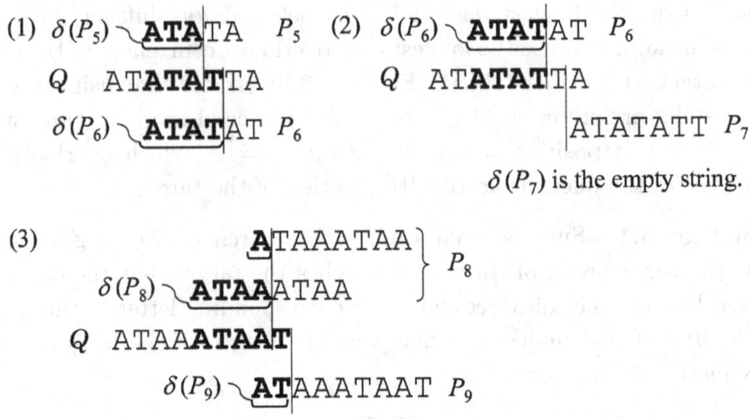

(1) $\delta(P_5)$ ⌐**ATA**TA P_5 (2) $\delta(P_6)$ ⌐**ATAT**AT P_6
    Q   A**TATAT**TA   Q   ATATA**TT**A
    $\delta(P_6)$ ⌐**ATAT**AT P_6 ATATATT P_7

                                              $\delta(P_7)$ is the empty string.

(3)   **A**TAAATAA ⎫
                                        ⎬ P_8
   $\delta(P_8)$ ⌐**ATAA**ATAA   ⎭
   Q   ATAA**ATAAT**
   $\delta(P_9)$ ⌐**AT**AAATAAT P_9

Fig. 5.5 Three examples illustrate how to compute the longest suffix-prefix match of P_{i+1} from P_i. Strings of bold letters show suffix-prefix matches.

hand. We will present an algorithm for this purpose and introduce some notations to describe the algorithm.

Definition 5.2 Let P_i be the i-mer prefix of the given query Q. Let $\delta(P_i)$ denote the longest suffix-prefix match of P_i, and let $|\delta(P_i)|$ denote its length.

We will compute $|\delta(P_i)|$ ($i = 1, \ldots, n$) inductively. First, P_1 is the 1-mer prefix and has only the empty proper suffix, indicating that $\delta(P_1)$ is the empty string. In the inductive step, we calculate $\delta(P_{i+1})$ by assuming that $\delta(P_i)$ is available. Figure 5.5 illustrates three typical cases.

(1) $\delta(P_6)$ is $\delta(P_5)$ appended with T ($= Q[5]$). This is the ideal case in which $\delta(P_{i+1})$ is obtained from $\delta(P_i)$ immediately.
(2) Concatenating $\delta(P_6)$ and T ($= Q[6]$) does not yield $\delta(P_7)$ because $\delta(P_7)$ is the empty string.
(3) Although $\delta(P_8)$ appended with T ($= Q[8]$) is not equal to $\delta(P_9)$, concatenating A, a suffix-prefix match of P_8, and T yields $\delta(P_9)$.

We will prove that a suffix-prefix match of P_i appended with $Q[i]$ gives $\delta(P_{i+1})$. For enumerating all suffix-prefix matches of P_i from longer ones, we will also show a systematic method based on the observation that, since $\delta(P_i)$ is a proper suffix of P_i, $\delta(\delta(P_i))$ is also a proper suffix of P_i. Let $\delta^k(P_i)$ denote the string obtained by k applications of δ to P_i.

Theorem 5.1 Any suffix-prefix match of P_i is equal to $\delta^k(P_i)$ for some k.

Proof. Otherwise, suppose that a j-mer suffix-prefix match is not of the form. There exists $k(\geq 1)$ such that $|\delta^k(P_i)| > j > |\delta^{k+1}(P_i)|$ because $|\delta^k(P_i)|(k = 1, 2, \ldots)$ decreases monotonically and reaches zero eventually. In this case, the j-mer must be a suffix-prefix match of $\delta^k(P_i)$ and is also longer than $\delta^{k+1}(P_i)$, the longest suffix-prefix match of $\delta^k(P_i)$. This is a contradiction. □

Since $\delta(P_{i+1})$ is longest, it is sufficient to compute the largest j ($< i$) such that $\delta(P_{i+1})$ is the j-mer suffix-prefix match of P_i appended with $Q[i]$, as shown in Figure 5.6.

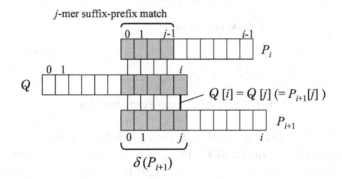

Fig. 5.6 Computing $\delta(P_{i+1})$ from a suffix-prefix match of P_i.

Algorithm 5.2 Assume that $|\delta(P_i)|$ is available by the inductive hypothesis. Perform the following steps to compute $|\delta(P_{i+1})|$:

(1) $k = 1$.
(2) Let $j = |\delta^k(P_i)|$.
(3) If $Q[i] = Q[j]$, j must be the largest j ($< i$) such that $\delta(P_{i+1})$ is the j-mer suffix-prefix match of P_i followed by $Q[i]$. Therefore, set $|\delta(P_{i+1})| = j + 1$, and exit.
(4) If $Q[i] \neq Q[j]$ and $j = 0$, $\delta(P_{i+1})$ is empty. Set $|\delta(P_{i+1})| = 0$, and exit.
(5) Otherwise, increment k and go to Step (2).

The program in Figure 5.7 implements the Algorithm 5.2. `delta[i]` stores $|\delta(P_i)|$. To compute the table `delta`, `j` is used to contain a temporary

```
public static void kmpSearch( int[] target, int[] query ) {
    int targetLen = target.length;
    int queryLen  = query.length;
    // Compute the delta table.
    int[] delta = new int[queryLen+1];
    delta[0] = -1;
    for(int i = 0, j = -1; i < queryLen; i++, j++, delta[i]=j){
        while( (0 <= j) && query[i] != query[j] )  j = delta[j];
    }
    // Search the target for the query using the KMP shift rule.
    for(int i = 0, j = 0; (i < targetLen) && (j < queryLen); i++, j++){
        if(target[i] != query[j]){ // 0 <= j at this moment.
            j = delta[j];
            while( (0 <= j) && (target[i] != query[j]) )  j = delta[j];
        }else
            if(j == queryLen-1){   // Print positions of query occurrences.
                System.out.print(i-j+" ");
                i++;
                j = delta[queryLen];
            }
    }
}
```

Fig. 5.7 Knuth-Morris-Pratt algorithm.

value of $|\delta^k(P_i)|$. The upper table in Figure 5.8 shows how the `delta` table is generated for the query ATATATT. The lower table in the figure shows how the KMP shift rule searches the target for the query using the `delta` table. Finally, we will show that the time complexity of the KMP algorithm is $O(n + m)$, where n and m denote the lengths of the target and query respectively. We will estimate the number of character comparisons that comprise the dominant steps in the algorithm.

Theorem 5.2 The KMP algorithm makes at most $2(n + m)$ character comparisons.

Proof. First, we will estimate how many character comparisons "query[i] != query[j]" are made during the calculation of the `delta` table. Since the program executes "j = delta[j]" or "j++" after the character comparison, the number of character comparisons is bounded by the number of times that the value of j is updated. In the first for-loop, each of variables i and j is incremented m times. Statement "j = delta[j]" decreases j by at least one, and the number of steps executing the statement must be at most m because j is incremented m times, and j is required to be no less than zero after the first for-loop. Therefore, j is updated at

i	j	0	1	2	3	4	5	6	7
1		A							
	0	A	T	A	T	A	T	T	
1		A	T						
	0	A	T	A	T	A	T	T	
	-1		A	T	A	T	A	T	T
2	0	A	T						
			A	T	A	T	A	T	T
3		A	T	A					
	1		A	T	A	T	A	T	T
4		A	T	A	T				
	2			A	T	A	T	A	T	T

i	j	0	1	2	3	4	5	6	7
5		A	T	A	T	A			
	3		A	T	A	T	A	T	T
6		A	T	A	T	A	T		
	4		A	T	A	T	A	T	T
6		A	T	A	T	A	T	T	
	4		A	T	A	T	A	T	T
	2			A	T	A	T	A	T	T
	0					A	T	A	T	A	T	T
	-1						A	T	A	T	A	T	T
7		A	T	A	T	A	T	T	
	0					A	T	A	T	A	T	T

			0	1	2	3	4	5	6	7	8	9	10	11	12	13	14	15	16	17	18
i	j	delta	A	T	A	T	A	T	A	T	T	A	A	T	A	T	A	T	T
0, ..., 6	0, ..., 6	delta[6]=4	A	T	A	T	A	T	T											
6, 7, 8	4, 5, 6	delta[6]=4					A	T	A	T	A	T	T							
8, 9, 10, 11	4, 5, 6, 7	delta[7]=0							A	T	A	T	A	T	T					
11, 12	0, 1	delta[1]=0												A	T					
12,..., 18	0, ..., 6														A	T	A	T	A	T	T

Fig. 5.8 The upper table shows how the KMP algorithm calculates the `delta` table for `ATATATT`. Bold letters indicate i-mer prefixes and their suffix-prefix matches that are identical to prefixes of the query. It also illustrates the temporary states of shifted queries. The lower table illustrates how the KMP algorithm searches the target string for query `ATATATT`. Only bold letters in the query examples are compared with the target. Letters in gray boxes disagree with their counterparts, and the queries are shifted.

most $2m$ times.

Second, we show that during applications of the KMP shift rule, the number of character comparisons by "`target[i] != query[j]`" is bounded by the number of times that `j` is updated, because the program performs "`j = delta[j]`" or "`j++`" after one character comparison. In the second for-loop, each of the variables `i` and `j` is incremented n times. The number of times `j` decreases by "`j = delta[j]`" is at most n, since `j` is incremented n times and $j \geq 0$ after the second for-loop. Overall, at most $2n$ character comparisons are made.

In total, at most $2(n+m)$ character comparisons are made. □

5.4 Bad Character Heuristics

Since nucleotide sequences consist of four letters, the query and the target strings are likely to share short substrings. In the KMP algorithm, these common substrings are utilized to shift the query effectively. In contrast, protein sequences are composed of 20 amino acids residues, and hence, the query and the target are less likely to share common substrings. For example, consider protein sequence

MVLSPADKTNVKAAWGKVGAH,

which are the first 21 amino acids residues in the protein sequence of Homo sapiens hemoglobin alpha 1 (accession number, NM_000558). Let us search this protein sequence for the last seven letters WGKVGAH. The KMP-algorithm will shift the query one by one whenever it identifies a mismatch, because all letters in the query are different, and no substrings occur anywhere else. The upper table of Figure 5.9 illustrates how the algorithm scans the target for the query.

For a larger number of letters in sequences, the probability of hitting a *nonexistent character* that appears in the target but does not appear in the query, increases. Identification of a nonexistent character makes it pointless to search the query for the character, thereby allowing us to shift the query entirely. In contrast, if a character in the target occurs in the query, the query is shifted to the right so that the "rightmost" occurrence in the query is aligned with the character in the target. Note that this shift is minimum so as to enumerate all occurrences of the query. This idea is called the *bad character heuristics* and was devised in the Boyer-Moore algorithm [10].

The lower picture in Figure 5.9 shows how the heuristics process the running example. Bold letters in the query are compared with their counterparts in the target, and letter in gray boxes indicate mismatches. In the lower picture, the heuristics attempt to align the query with the target from the beginning. Pairs of characters are scanned from right to left; immediately the last pair disagrees. Since D does not occur in the query, it does not make sense to continue scanning the query. The query is therefore shifted entirely, and the comparison starts from the 13th position in the target. The last pair is inconsistent again, but A occurs in the query, and hence the query is shifted by one to restart the attempt at alignment from the last position. Although W in the 14th position disagrees with the last letter of the query, H, W occurs in the 0th position of the query. Therefore, two occurrences of W are aligned, and a pairwise letter comparison from

i	j	0	1	2	3	4	5	6	7	8	9	10	11	12	13	14	15	16	17	18	19	20
		M	V	L	S	P	A	D	K	T	N	V	K	A	A	W	G	K	V	G	A	H
0	0	**W**	G	K	V	G	A	H														
1	0		**W**	G	K	V	G	A	H													
2	0			**W**	G	K	V	G	A	H												

i	j	0	1	2	3	4	5	6	7	8	9	10	11	12	13	14	15	16	17	18	19	20
		M	V	L	S	P	A	D	K	T	N	V	K	A	A	W	G	K	V	G	A	H
6	6	W	G	K	V	G	A	**H**														
13	6								W	G	K	V	G	A	**H**							
14	6									W	G	K	V	G	A	**H**						
20, 19, ..., 14	6, 5, ..., 0																W	G	K	V	G	A	H

Fig. 5.9 The upper and lower pictures show how the KMP algorithm and the bad character heuristics process the target and the query, respectively.

right to left identifies the query in the target.

Figure 5.10 illustrates a Java program that implements the bad character heuristics. Before scanning the target and the query, the algorithm preprocesses the query and generates the shift table describing how far the target should be shifted to the right for each character in the target. The character is associated with queryLen if it is nonexistent, and queryLen-1-k otherwise, where k is the position of the character in the query. For example, the shift table for WGKVGAH is

	W	K	V	G	A	H	other letters
shift	6	4	3	2	1	0	7

Note that G occurs twice, but only the rightmost one is considered.

Subsequently, the program in Figure 5.10 scans the target and the query string simultaneously from right to left by using two variables, i and j, that indicate indexes for the two respective strings. The length of the query is denoted by queryLen. Suppose that the j-th character in the query mismatches the i-th in the target. The algorithm sets j to queryLen-1 in order to restart searching the query from the rightmost position. Accordingly, i must be updated.

If the i-th letter does not occur in the query, we skip the query entirely, as illustrated in Figures 5.11 (A) and (B). If the current i-th letter appears

```
public static void badCharHeuristics( int[] target, int[] query ) {
    int targetLen = target.length;
    int queryLen  = query.length;
    int SIZE = 256; // Declare the number of letters to use.
    int[] shift = new int[SIZE+1];
    // Generate the shift table.
    // Initially suppose that no letters appear in the query.
    for(int k = 0; k <= SIZE; k++) shift[k]=queryLen;
    // Associate the position of the rightmost concurrence with each letter.
    for(int k = 0; k < queryLen-1; k++)
        if(query[k] < SIZE) shift[query[k]] = queryLen-1-k; else return;
    // Search the target for the query.
    for(int i = queryLen-1, j = queryLen-1; i < targetLen; i--, j--){
        if(j == -1){ // Print positions of query occurrences.
            System.out.print(i+1+" ");
            i = i+1+queryLen;
            j = queryLen-1;
        }
        while(i < targetLen && target[i] != query[j]){
            int shiftLen = shift[target[i]];
            if(queryLen-j > shiftLen)
                i = i+queryLen-j;  else  i = i+shiftLen;
            j = queryLen-1;  // Search the query from the rightmost again.
        }
    }
    System.out.println();
}
```

Fig. 5.10 Bad character heuristics algorithm in the Boyer-Moore algorithm.

in the query, it is aligned with its rightmost occurrence in the query. This alignment is ensured by adding 6 (= shift[W]) to i as shown in Figures 5.11 (C) and (D).

However, this adjustment is not always successful as presented in Figure 5.11 (E). Four letters are scanned, and G in the target mismatches with its counterpart. Figure 5.11 (F) shows that in order to align G, the addition of 2 (= shift[G]) to i shifts the query to the "left" by one. However, this operation causes an immediate disagreement between A and H. Therefore, we try to align A by shifting the query to the right by one (Figure 5.11 (G)), leading to the same state shown in Figure 5.11 (E). Consequently, the process alternates between the states in Figures 5.11 (E) and (F) forever.

In order to break the infinite loop, we move the query to the right by just one character, as shown in Figures 5.11 (E) and (H). One may be concerned about missing a query occurrence because of this operation. However, the bad character heuristics shifts the query by the minimum length to avoid

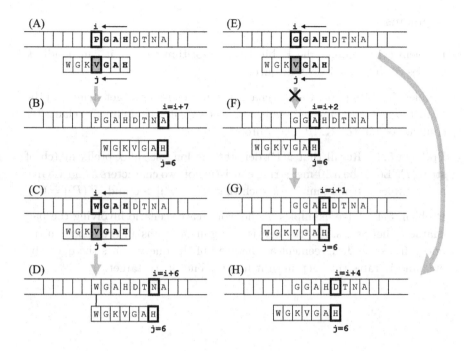

Fig. 5.11 How many characters should be shifted in order to devise the bad character heuristics?

this problem.

The original Boyer-Moore algorithm incorporates another rule called *good suffix heuristics*, although we do not discuss the idea in this book. See the details of these heuristics in [23, 35]. The bad character heuristics is typically much faster than the KMP algorithm when many distinct letters are used in strings such as ordinary texts. However, in the worst case, the bad character heuristics may perform $(n - m + 1)m$ character comparisons, where n and m are the lengths of the target and the query. In other words, there are cases when the heuristic argument scans all the letters in the query, finds the disagreement at the head of the query, shifts the query by one, and iterates this step until it hits the end of the target. We will leave this question to the reader.

Problems

Problem 5.1 Extend the Rabin-Karp algorithm so that it can handle k queries of the same length at once.

Problem 5.2 Suppose that all characters in the n-mer target string and the m-mer query string are the same, e.g., A. Estimate the number of character comparisons in the KMP algorithm.

Problem 5.3 Recall that $\delta(P)$ denotes the longest suffix-prefix match of string P. Let P be a 10-mer string consisting of two characters A and T. Give an example of P maximizing k such that $|\delta^{k-1}(P)| > 0$ and $|\delta^k(P)| = 0$.

Problem 5.4 Give examples of the worst-case scenario involving the bad character heuristics; i.e., the heuristic argument scans all the letters in the query, finds the disagreement at the head of the query, shifts the query by one, and iterates this step until it hits the end of the target.

Chapter 6

Approximate String Search

So far, we have discussed algorithms that efficiently search the target string for exact copies of the query string. In reality, however, there are compelling demands when looking for partial query matches to find those functionally meaningful sequences that are likely to be conserved across diverse genomes.

For example, Figure 6.1 compares the genomes of eight species near the first exon of *HOXA1*, a homeobox gene. In vertebrates, homeobox genes, which comprise a class of transcription factors, are expressed during embryonic development. *HOXA1* is located on human chromosome 7 and encodes a DNA-binding transcription factor that may control differentiation, morphogenesis, and gene expression. Homeobox genes are essential genes that are positively selected. Despite this, Figure 6.1 shows many differences between the sequences of species that diverged very long ago. In the presence of these mutations, it is impractical to look for exact matches, and is better to search for partial matches of queries in a process called an *approximate string search*.

Partial matches are often represented by *alignments*. The two lower pictures in Figure 6.1 are sample alignments between human and zebrafish sequences, and between fugu (Japanese puffer fish) and zebrafish sequences. A vertical bar (|) denotes the match of two characters in the column, while "x" indicates a mismatch. A horizontal bar "-," called the *gap symbol*, is inserted into a string if no character in the string corresponds to its counterpart. Note that the alignment between the human and zebrafish sequences has more mismatches than the alignment between fugu and zebrafish.

In this chapter, we present several algorithms for calculating alignments that allow partial matches. All of the algorithms are highly accurate in detecting partially matched regions of a lower identity at the expense of computational efficiency. In fact, these algorithms run in $O(mn)$ time when

Human	GACAATGCAAGAATGAACTCCTTCCTGGAATACCC---CATA
Chimp	GACAATGCAAGAATGAACTCCTTCCTGGAATACCC---CATA
Mouse	GACAATGCAAGAATGAACTCCTT**T**CTGGAATACCC---CAT**C**
Rat	GACAATGCAAGAATGAACTCCTT**T**CTGGAATACCC---CAT**C**
Dog	GACAATGCAAGAATGA**G**CTCCTTCCTGGAATACCC---CAT**C**
Chicken	GACAAT**A**CTAG**G**ATGAACTCCTT C**TTA**GAGTAT**GC**---**A**AT**T**
Fugu	-ACAATGC**CAC**AATGAG**C**AG CTT C**TTA**GAT**TA**CTC---**TGTG**
Zebrafish	GA**AG**ATG**ACAC**AATGAG**C**ACATTC**TTA**GAT**TTT**CGTCCATA

Amino Acids	M	N	S	F	L	E	Y	P	I
Synonymous mutations	ATGAAC	TCC	TTC	CTG	GAA	TAC	CCC	ATA	
		AGC	TTT	TTA		GAG	TAT		ATC
									ATT
		S	T			D	F		V
		AGC	ACA			GAT	TTT		GTG

```
Human     GACAATGCAAGAATGAACTCCTTCCTGGAATACCC---CATA
          ||xx|||xx|x|||||x|x|x|||x|x||x|xxx|    ||||
Zebrafish GAAGATGACACAATGAGCACATTCTTAGATTTTCGTCCATA

Fugu      -ACAATGCCACAATGAGCAGCTTCTTAGATTACTC---TGTG
           |xx|||x|||||||||||xx|||||||||||xx||   xx|x
Zebrafish GAAGATGACACAATGAGCACATTCTTAGATTTTCGTCCATA
```

Fig. 6.1 The upper picture compares the genomes of eight species at the first exon of *HOXA1*, a homeobox gene. Each bold nucleotide indicates a difference from its counterpart in the human genome. In the middle, the amino-acid sequence is translated from the protein-coding region. Note that most mutations are synonymous and due to the redundancy of the code, no change in the protein occurs. Two alignments are shown at the bottom. The accession number of *HOXA1* is NM_005522. The build number of the human genome is hg17, which was assigned in May 2004. The human nucleotide sequence ranges from 26,908,730 to 26,908,768 in the "minus" strand of human chromosome 7.

two strings of length m and n are compared. Therefore, they are typically too slow to compare long genomic sequences. Methods of accelerating the approximate string search are presented in the next chapter.

6.1 Edit Operations and Alignments

Finding an approximate match between two similar strings can be viewed as a series of operations editing one string with another string. For example, let us consider identifying the similarity between ATCGAT and ATGCGT by editing the former string to the other. Basic edit operations involve substituting one character for another, inserting a new character into a string, and deleting one character. By applying a series of this fundamental edit operation, one can transform one string into any other string. Figure 6.2 (A) presents three series of edits that yield ATGCGT from ATCGAT. Edit operations should be nonredundant. It would be pointless to substitute C for G and then substitute G for the original character C, or to delete and then reinsert the same character.

Describing a series of several edit operations can be lengthy, as seen in Figure 6.2 (A). Alignments are useful for showing which characters are edited quickly.

Definition 6.1 Given two input strings S_1 and S_2, insert gap symbols "-" into both strings so that the resulting two strings are of equal length and two characters in any position must not create gaps at the same time to avoid the unnecessary insertions of gaps. A *global alignment* of two input strings S_1 and S_2 is a three-row matrix such that the first and third rows store the two resulting strings after gaps are inserted, and the second row expresses the correspondence between two characters in each column; i.e., the match is represented by "|," the mismatch by "x," and a pair with a gap by a space. A *local alignment* of S_1 and S_2 represents a global alignment of substrings of S_1 and S_2.

In a global alignment, all the characters in the two input strings must be involved in the alignment, but this strict condition is not required in local alignments. We will discuss the benefit of local alignments later in this chapter, but for the present, we will discuss global alignments. For example, the global alignment in the middle of the left part in Figure 6.2 (B) displays three substitutions performed by aligning three pairs of changed letters and putting the mismatch symbol x between each pair. The gap symbol "-" is used to indicate the application of an insertion and deletion, as illustrated in Figure 6.2 (B).

Given a global alignment, the degree of similarity between the two strings can be measured by assigning appropriate scores to the individual columns. The sum of these scores estimates the overall similarity.

(A) Series of edit operations

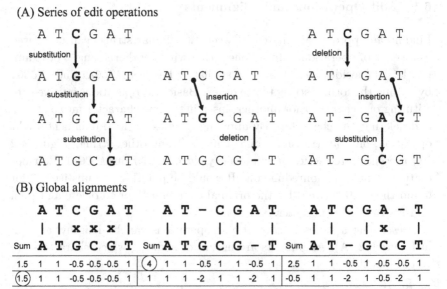

(B) Global alignments

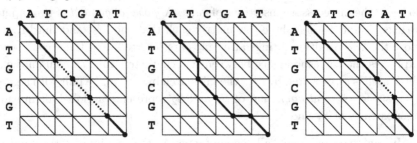

The gap penalty is set to -0.5 in the upper row and -2 in the lower row.
The best scores in the two rows are circled.

(C) Edit graphs

Fig. 6.2 (A) These three series of edit operations transform ATCGAT into ATGCGT. The characters manipulated by the edit operations are shown in bold. (B) The global alignments outline the corresponding edit operation series in a concise manner. The scores beneath each column indicate the degrees of agreement between characters. The match score, mismatch penalty, and gap penalty are set to 1, -0.5, and -0.5, respectively, in the upper row, while the gap penalty is changed to -2 in the lower row. (C) In the three edit graphs, the bold paths illustrate the series of edit operations and alignments. The dotted bold edges indicate mismatches.

Definition 6.2 In a global alignment, if two letters in a column agree, we assign the *match score*, i.e., 1.0, to the column; otherwise, we assign the *mismatch penalty* (score), i.e., -0.5, to the column. If one of the two letters in a column is the gap symbol, we assign the *gap penalty* (score), i.e., -0.5, to the column. The sum of the scores in all columns is called the *alignment score* or simply the *score* of the alignment.

The choice of scores requires careful consideration. For example, mutations are likely to be observed more frequently than insertions and deletions if the genomes of divergent species are compared. In this case, we must penalize gaps more severely than mismatches by setting the gap penalty to, say -2, which is much greater than the mismatch penalty, -0.5. This alternative measurement is also presented in Figure 6.2 (B). Note that the best alignment with the maximum score changes with the penalty values selected.

6.2 Edit Graphs and Dynamic Programming

To compute the best alignment with the maximum score, we will first introduce a data structure called an edit graph.

Definition 6.3 An *edit graph* is a two-dimensional matrix such that the two input strings are displayed above the top horizontal row and beside the left vertical column, respectively. A path in an edit graph represents an alignment, but in order to avoid useless edit operations, we impose the constraint on paths so that it is allowable to move from a node in three directions, to the right, to the lower-right, and to the bottom. A path that satisfies this requirement is called an *alignment path*. An alignment path from the top-leftmost node to the bottom-rightmost node is called *global*, while an alignment path from any node to some other node is called *local*. The *score* of an alignment path is the score of its corresponding alignment.

Figure 6.2 (C) illustrates three paths that correspond to the above alignments. Observe that a move to the right results in the deletion of the letter in the top horizontal string, a move along the diagonal denotes the match or mismatch of two characters, and a move to the bottom indicates the insertion of a letter. It becomes obvious that any global (or local) alignment can be represented by a global (local) alignment path in the edit graph, and that any alignment path denotes an alignment. The rigorous proof can be ascertained by the induction on the length of the alignment and the alignment path.

The score of an alignment path can be computed directly without reconstructing its corresponding alignment because moving to the right or to

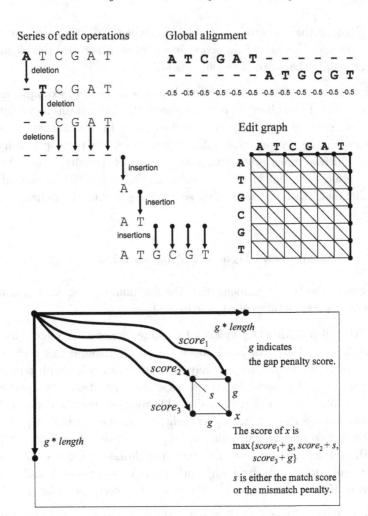

Fig. 6.3 The upper half shows another example of an edit operation series, its alignment, and its edit graph. The lower half illustrates the concept of dynamic programming for calculating optimal paths with the maximum score.

the bottom simply adds the gap penalty to the overall score, and moving diagonally adds the match score or the mismatch penalty. Computing the best global alignment is equivalent to calculating the best global alignment path with the maximum score. However, this is a nontrivial task because there may be numerous alignment paths in an edit graph. We may have to consider pathological alignment paths, such as illustrated in the upper half of Figure 6.3. The number of alignment paths can easily lead to a combinatorial explosion because the number must be at least 3^n, where n is the length of the shorter input string. Consider alignment paths that visit all n nodes on the diagonal starting at the top-leftmost node. The number of such alignment paths equals 3^n because three choices are involved in moving diagonally from one node to the next: to the right and then down, to the lower right along the diagonal, and down and then to the right.

Fortunately, to find the best alignment path with the maximum score, we can ignore many irrelevant paths and gain a performance benefit by using the dynamic programming approach, which computes the best global alignment path from the top-leftmost node to each node iteratively. The lower half of Figure 6.3 illustrates this concept. The main idea is that the score of the best alignment path to node x must be determined by the three best alignment paths to its three neighboring nodes that are able to access x directly. It becomes obvious that the score of the best alignment path to x is the maximum score of the best alignment path to the upper node plus the gap penalty g, the score of the best alignment path to the upper-left node plus the match score or the mismatch penalty score, and the score of the best alignment path to the left node plus g. Otherwise, we have an immediate contradiction because of the choice of the best alignment path to x. Care must be taken to calculate the scores of nodes in the top line and in the left line because each has only one previous node, not three.

To implement the dynamic programming approach, we need to scan the two-dimensional matrix in some way so that the previous nodes of the current node are processed and associated with their maximum scores. One typical scanning strategy is to conduct a step-by-step sweep of each column from top to bottom and iterate this process to the columns from left to right.

6.3 Needleman-Wunsch Algorithm

The Needleman-Wunsch algorithm [70] utilizes the dynamic programming approach outlined in the previous section to compute the best global alignment. It generates a two-dimensional matrix of real numbers called the **score** such that the number of columns equals one plus the length of one string and the number of rows is also one plus the length of the other input string. The matrix is used to store the maximum scores of the best alignment paths to individual nodes.

The algorithm initializes the scores of the boundary nodes in the top and leftmost lines so that the score equals the gap penalty times the distance from the top leftmost node to the current node because the best alignment paths to these boundary nodes must be straight paths from the top leftmost node. Then, it calculates the score of the node score[x][y] by consulting the scores of three previous nodes, score[x-1][y-1], score[x-1][y], score[x][y-1]. It adds the match score, mismatch penalty, or gap penalty to these scores appropriately, and takes the highest of the three values to output the alignment score of the path to the node. The upper half of Figure 6.5 implements these procedures. The program sweeps columns from left to right and then scans elements in one column from top to bottom.

Figure 6.4 shows an example of the score matrix for computing the alignment between ATCGAT and ATGCGT. The first half of the program computes and stores the score of the best alignment path to each node, but discards the best alignment paths to individual nodes. One might be concerned about this, but it presents no problem because the best alignment path to a node can be restored from the maximum alignment score assigned to the node. For example, consider the maximum value 4 in the bottom rightmost corner node in Figure 6.4. In the matrix, score[6][6] has 4. The value 4 can be only computed from 3 in the upper-left node score[5][5] by confirming the match of the last characters Ts in the two input strings, i.e., 4 is equal to 3 plus the match score 1. Next, the value

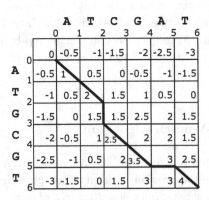

Fig. 6.4 The Needleman-Wunsch algorithm computes the maximum scores of the best alignment paths and puts them into the score array. The match score, mismatch penalty, and gap penalty are set to 1, -0.5, and -0.5, respectively. The bold path denotes the best alignment path, which has the maximum score 4.

```
final static double matchScore      = 1;
final static double mismatchPenalty = -0.5;
final static double gapPenalty      = -0.5;
public static void NeedlemanWunsch(int[] stringX, int[] stringY){
    // Initialize constants.
    int stringXlen = stringX.length; int stringYlen = stringY.length;
    double[][] score = new double[stringXlen+1][stringYlen+1];
    // Initialize the score array.
    for(int x=0; x < stringXlen+1; x++) score[x][0] = gapPenalty*x;
    for(int y=0; y < stringYlen+1; y++) score[0][y] = gapPenalty*y;
    // Inductive steps.
    double oneScore;
    for(int x=1; x < stringXlen+1; x++){
        for(int y=1; y < stringYlen+1; y++){
            if(stringX[x-1] == stringY[y-1]) oneScore = matchScore;
            else oneScore = mismatchPenalty;
            score[x][y] = max( score[x-1][y-1] + oneScore,
                               score[x-1][y]   + gapPenalty,
                               score[x][y-1]   + gapPenalty); }
    }
    // The traceback procedure for restoring the alignment
    // from the bottom-rightmost to the top-leftmost.
    int[][] alignment = new int[stringXlen+stringYlen+1][2]; int i = 0; int x,y;
    for(x = stringX.length, y = stringY.length; !(x == 0 && y == 0); i++){
        if(0 < x && 0 < y && stringX[x-1] == stringY[y-1])
            oneScore = matchScore;
        else oneScore = mismatchPenalty;
        if(0 < x && 0 < y && score[x][y] == score[x-1][y-1] + oneScore){
            alignment[i][0] = stringX[x-1]; alignment[i][1] = stringY[y-1];
            x = x-1; y = y-1;
        }else if(0 < x && 0 <= y && score[x][y] == score[x-1][y] + gapPenalty){
            alignment[i][0] = stringX[x-1]; alignment[i][1] = -1; x = x-1;
            // -1 means a gap.
        }else if(0 <= x && 0 < y && score[x][y] == score[x][y-1] + gapPenalty){
            alignment[i][0] = -1; alignment[i][1] = stringY[y-1]; y = y-1; }
    }
    printAlignment(alignment, i-1);
}
public static void printAlignment(int[][] a, int lastPos){
    for(int j=lastPos; 0<=j; j--) System.out.print(int2char(a[j][0])+" ");
    System.out.println();
    for(int j=lastPos; 0<=j; j--)
        if(a[j][0] == -1 || a[j][1] == -1) System.out.print("  ");
        else if(a[j][0] == a[j][1]) System.out.print("| ");
        else System.out.print("x ");
    System.out.println();
    for(int j=lastPos; 0<=j; j--) System.out.print(int2char(a[j][1])+" ");
    System.out.println();
}
public static char int2char(int i){ if(i == -1) return '-'; else return (char)i; }
```

Fig. 6.5 The Needleman Wunsch algorithm.

3 must be obtained from 3.5 in the left node `score[4][5]` by adding the negative gap penalty, i.e., 3 = 3.5 + (-0.5).

This task is called the *traceback* procedure. The lower half of Figure 6.5 implements the traceback procedure. It first creates an array for storing the restored alignment. Starting from the bottom-rightmost node, the traceback procedure examines the scores of the three previous nodes and determines which one will provide the score of the current node. When multiple nodes are candidates, any one can be selected.

Finally, let us analyze the worst-case scenario involving the computational complexity of the Needleman-Wunsch algorithm. Let m and n denote the lengths of the two input strings `stringX` and `stringY`. The first half of the program scans $(m+1)(n+1)$ individual elements in array `score`, and each visit executes several statements that must be performed in a constant time. The traceback procedure in the lower half visits at most $m + n + 1$ nodes and each visit requires a constant time operation. Therefore, the worst-case time complexity is $O(mn)$.

6.4 Smith-Waterman Algorithm for Computing Local Alignments

Global alignment paths must start at the top-leftmost node and end at the bottom-rightmost node. This requirement may hinder our finding a strong similarity between the substrings of input strings. To identify such local similarity, it is more suitable to look for local alignment paths that can start from any node and end at any node.

For example, Figure 6.6 presents the best global alignment with the maximum score of 2.5 and the best local alignment with the maximum score of 4 when the match score, mismatch penalty, and gap penalty are set to 1, -0.5, and -0.5. The score of the local alignment is better than that of the global alignment, because all characters in the input strings are not necessarily involved in local alignments, making it possible to ignore useless pairs that do not contribute to increasing the overall score. Note that using local alignments reveals the consecutive four letter matches, which is overlooked when using the best global alignment.

In 1981, Smith and Waterman made three slight modifications to the Needleman-Wunsch algorithm to derive a more efficient algorithm for computing the best local alignment path [97]. Figure 6.8 presents a program that implements the Smith-Waterman algorithm, and its operation is illus-

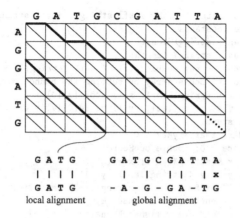

Fig. 6.6 Comparison between global and local alignments.

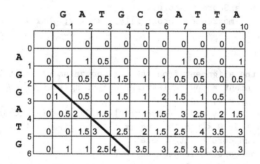

Fig. 6.7 Scores computed using the Smith-Waterman algorithm and the best local alignment path.

trated in Figure 6.7.

First, the scores of the elements in the top row and in the leftmost column are set to zero, i.e., score[x][0] = 0 and score[0][y] for all possible values of x and y. This change allows a local alignment path to start from any node on the two boundaries. Second, to allow a local alignment path to begin at any node, the score of any present node can be set to zero if the values of the three previous nodes are negative, which has the effect of discarding the best alignment paths with negative scores; therefore, we are able to restart a new local alignment path from the present node. Third, the traceback procedure identifies the node with the maximum

```
public static void SmithWaterman( int[] stringX, int[] stringY){
    // Initialize constants.
    int stringXlen = stringX.length;
    int stringYlen = stringY.length;
    double[][] score = new double[stringXlen+1][stringYlen+1];
    // Initialize the score array by setting the boundary elements to 0.
    for(int x=0; x < stringXlen+1; x++) score[x][0] = 0;
    for(int y=0; y < stringYlen+1; y++) score[0][y] = 0;
    // Inductive steps.
    int xMax, yMax; xMax = yMax = 0;
    double maxScore = 0; double oneScore;
    for(int x=1; x < stringXlen+1; x++){
        for(int y=1; y < stringYlen+1; y++){
            if(0 < x && 0 < y && stringX[x-1] == stringY[y-1])
                oneScore = matchScore;
            else oneScore = mismatchPenalty;
            score[x][y] = max( 0,  // This allows generation of a new path.
                            score[x-1][y-1] + oneScore,
                            score[x-1][y]   + gapPenalty,
                            score[x][y-1]   + gapPenalty);
            if(maxScore < score[x][y]){
                maxScore = score[x][y]; xMax=x; yMax=y;
            }
        }
    }
    // Restore the alignment from the maximum value until it hits zero.
    int[][] alignment = new int[stringXlen+stringYlen][2];
    int i = 0; int x,y;
    for(x = xMax, y = yMax; !(score[x][y] == 0); i++){ // Changed to hit the zero.
        if(0 < x && 0 < y && stringX[x-1] == stringY[y-1])
            oneScore = matchScore;
        else oneScore = mismatchPenalty;
        if(0 < x && 0 < y && score[x][y] == score[x-1][y-1] + oneScore){
            alignment[i][0] = stringX[x-1]; alignment[i][1] = stringY[y-1];
            x = x-1; y = y-1;
        }else if(0 < x && 0 <= y && score[x][y] == score[x-1][y] + gapPenalty){
            alignment[i][0] = stringX[x-1]; alignment[i][1] = -1; x = x-1;
        }else if(0 <= x && 0 < y && score[x][y] == score[x][y-1] + gapPenalty){
            alignment[i][0] = -1;  alignment[i][1] = stringY[y-1]; y = y-1; }
    }
    printAlignment(alignment, i-1);
}
```

Fig. 6.8 Smith-Waterman algorithm.

score, which may not be the bottom-rightmost node, and it starts restoring the best local alignment path from the identified node until it hits a node of score zero, which is not always the top-leftmost node.

6.5 Overlap Alignments

Here, we introduce another type of alignment called the overlap alignment, which plays a fundamental role in assembling millions of DNA fragments. Since the genome DNA is sheared at random, two DNA fragments that originate from the same position in the genome are seldom identical, but the suffix of one fragment is likely to be highly similar to the prefix of another fragment, or one fragment may be subsumed by another. Therefore, computing the global alignment of two DNA fragments is rarely practical, but finding suffix-prefix partial matches or a subsumption relationship is essential.

Definition 6.4 An *overlap alignment* of S_1 and S_2 is a global alignment of a suffix (or prefix) of S_1 and a prefix (or suffix) of S_2, a global alignment of a substring of S_1 and S_2, or a global alignment of S_1 and a substring of S_2. The latter two cases indicate that one string subsumes the other. An *overlap alignment path* in an edit graph starts from a node in the top row or the leftmost column and ends at a node in the bottom row or the rightmost column.

Figure 6.9 shows an example of an overlap alignment path. The global alignment of the same input strings contains many gaps and mismatches, and hence fails to detect a long stretch of matches involving a suffix of the horizontal input string that is also a prefix of the vertical input string.

An efficient algorithm for computing the best overlap alignment path with the maximum score is obtained by making two slight modifications to the Needleman-Wunsch algorithm. The lower half of Figure 6.10 illustrates this algorithm. First, to allow alignment paths to start from an element in the top row or in the leftmost column, the scores of nodes in these two boundaries are set to zero, which is the same modification as in deriving an algorithm for local alignments. The second modification of the traceback procedure is of special interest. We scan the rightmost column and the bottom row to find the node with the maximum score because the best overlap alignment must terminate at this node. Then, we restore the overlap alignment from the node until we hit the top row or the leftmost column. In Figure 6.10, this traceback step is omitted and left for the reader to complete. The upper half of Figure 6.10 shows the operation of this algorithm.

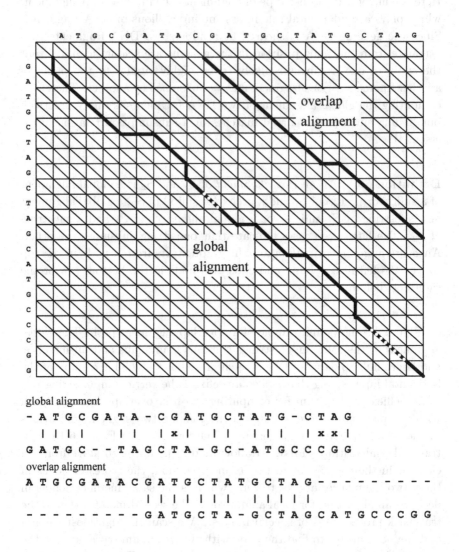

Fig. 6.9 An example of overlap alignment.

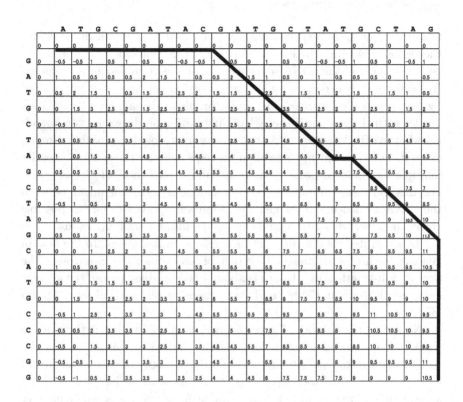

```
public static void overlapAlignment(int[] stringX, int[] stringY){
    // Initialize constants.
    (Identical to the Needleman-Wunsch algorithm.)
    // Initialize the score array.
    for(int x=0; x < stringXlen+1; x++) score[x][0] = 0;
    for(int y=0; y < stringYlen+1; y++) score[0][y] = 0;
    // Inductive steps.
    (Identical to the Needleman-Wunsch algorithm.)
    // The traceback procedure for restoring the best overlap alignment.
    (This part is left to the reader to complete.)
    printAlignment(alignment, i-1);
}
```

Fig. 6.10 Overlap alignment algorithm and its operation.

6.6 Alignment of cDNA Sequences with Genomes and Affine Gap Penalties

Once the genome of a species has been elucidated, one typically looks for the regions encoding the cDNA sequences of known genes. In vertebrate genomes, most of the cDNA sequences are divided into disjointed consecutive substrings called *exons*, which are coded in genomes, but interspaced by long gaps called *introns*. The major problem with aligning a cDNA sequence with the genome is the lack of information on the exon-exon boundaries in the cDNA sequence. A common computational approach to predict exons in a cDNA sequence is to align the sequence with the genome, identify multiple blocks of consecutive matches, and regard these blocks as exons. For this purpose, computing the best global or local alignment between the cDNA sequence and the genome of interest appears to be one possible choice; however, this approach is not successful in identifying exons.

To illustrate this difficulty, let us consider human genes. In the human genome, each gene contains an average of 8.8 exons and 7.8 introns. The average full-length cDNA length is about 1,500 bp (base pairs); the average exon length is about 170 bp; and the average intron size is about 5,419 bp [85].

Note that the number of gaps (introns) is fairly small, but the average length of one intron is much longer than that of one exon. Using the conventional *linear gap penalty* criterion that assigns the same penalty to each gap may allow too many gaps to be inserted as long as the total score is maximized, which is not suitable for detecting a small number of exons. Another drawback of the linear gap penalty is that the criterion penalizes long introns severely, making it difficult for the best global or local alignment path to incorporate long introns. We need a different standard of gap penalty for this application.

Given a block of consecutive l gaps, the linear gap penalty assigns a $g \times l$ gap penalty to the block, where g is one gap penalty. One commonly accepted modification of the linear gap penalty used to find a meaningful alignment of a cDNA sequence with genomes is the *affine gap penalty* in the form $h + g \times l$, where h is called the *startup gap penalty*; the startup gap penalty is given to a block of gaps only once. The startup penalty h should be much higher than one gap penalty g, e.g., by an order of magnitude. The modification penalizes the insertion of many gap blocks, which leads to the combination of short gap blocks into a single longer block to reduce startup gap penalties. In addition, as one gap penalty effect is reduced relatively,

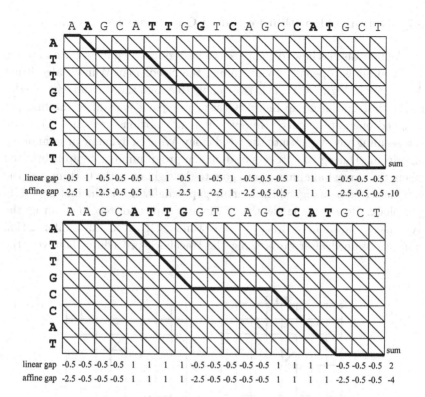

Fig. 6.11 In both edit graphs, the match score, mismatch penalty score, and gap penalty score are set to 1, -0.5, and -0.5, respectively. In addition, the startup gap penalty for the affine gap penalty is set to -2. In both graphs, the scores based on the linear and affine gap penalties are juxtaposed beneath the individual matrices. In each gap block, the startup gap penalty -2 is also assigned to the first gap. In the upper graph, the best global alignment path in terms of the linear gap penalty is represented by bold lines, while the bold lines in the lower graph denote the best global alignment path in terms of the affine gap penalty. Note that searching for the best global alignment path in the upper edit graph fails to identify two consecutive matches in the lower edit graph.

the difference in penalties between a short intron and a long intron becomes much less, making it less costly to insert long introns into alignments.

For example, see Figure 6.11. The upper edit graph presents the best global alignment path containing six blocks of continuous gaps. By contrast, the alignment path in the lower edit graph adopts an affine gap penalty and has only three blocks of contiguous gaps; the alignment path also involves two long matches.

6.7 Gotoh's Algorithm for Affine Gap Penalties

We now discuss how to compute the best global alignment path in terms of the affine gap penalty by extending the Needleman-Wunsch algorithm. As we saw in Figure 6.11, it is reasonable to assign the startup gap penalty to the first gap in a gap block. Other subsequent gaps after the first are assigned one gap penalty. However, it is not obvious how to incorporate this seemingly easy task into the Needleman-Wunsch dynamic programming algorithm. During the process of dynamic programming, a gap is inserted where the current alignment path goes to the right or to the bottom, and the gap must also be assigned the startup gap penalty if it is the first gap in its gap block. However, the identity of the first gap is not recorded during the dynamic programming process. One wonders if the missing information can be obtained afterwards using the traceback procedure; however, the

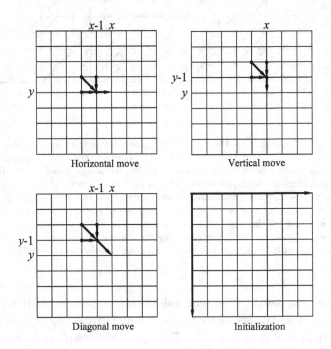

Fig. 6.12 The key concept behind Gotoh's algorithm is to consider which of the three cases was involved in the last move to the present node score[x][y], i.e., was it horizontal, vertical, or diagonal. The nodes in the top and leftmost lines must be initialized carefully.

missing information itself is essential to compute the best alignment.

We must memorize the information that will allow us to decide whether the current gap is the first one. In 1982, Osamu Gotoh stated that it was sufficient to record the scores of the best three paths such that the last move to the present node was the horizontal move from the left node, the vertical move from the upper node, or the diagonal move from the upper-left node [31]. See Figure 6.12.

Suppose that the last move to the current node score[x][y] was horizontal. If the move to the previous node score[x-1][y] was either vertical or diagonal, a new gap block would be created in the horizontal direction, and hence the startup gap penalty must be assigned to the last move in addition to one gap penalty. Otherwise, if the move to the previous was also horizontal, the current move is not the first gap, and therefore it would be sufficient to assign one gap penalty to the last move.

Similarly, when the last move is vertical, the startup gap penalty is added if the previous move is horizontal or diagonal, and only one gap penalty is enough if the previous move is also vertical. When the move is diagonal, no new gap blocks are created at the present node, indicating that there is no need to consider the startup gap penalty.

The program in Figure 6.13 implements this idea in its inductive steps from the middle of the program. The program uses three matrices, diagonal, horizontal, and vertical, in place of the single score matrix because the best scores must be kept for each of the three types of last move to the present node. In the initialization step, since only horizontal moves are possible for accessing any node in the top line, horizontal[x][0] is set to one gap penalty times the distance from the top-leftmost node to the current node, while diagonal[x][0] and vertical[x][0] are set to an extremely small value so that the best alignment path does not involve these values. Similarly, only vertical moves are allowed to approach any node in the leftmost column, and the initialization is performed accordingly.

The program then scans the three matrices. The strategy of scanning elements is the same as that of the Needleman-Wunsch algorithm, i.e., from the leftmost column to the right and from the top row to the bottom in each column. In the inductive step, the three values of diagonal, horizontal, and vertical are calculated by consulting the values of the previous three nodes, as shown in Figure 6.13. After all the best scores are calculated, the traceback procedure in Figure 6.14 starts at the bottom-rightmost node and selects the best score from among the three values in diagonal, horizontal, and vertical. Then, it iterates the step of moving

```
final static double matchScore        = 1;
final static double mismatchPenalty   = -0.5;
final static double gapPenalty        = -0.5;
final static double startupGapPenalty = -2;

public static void Gotoh(int[] stringX, int[] stringY){
    // Initialize constants.
    int stringXlen = stringX.length;
    int stringYlen = stringY.length;
    double lowestScoreSum = Math.min(startupGapPenalty,
            Math.min(mismatchPenalty, gapPenalty))*(stringXlen + stringYlen);
    double[][] diagonal   = new double[stringXlen+1][stringYlen+1];
    double[][] vertical   = new double[stringXlen+1][stringYlen+1];
    double[][] horizontal = new double[stringXlen+1][stringYlen+1];
    // Initialize the score array.
    for(int x=0; x < stringXlen+1; x++){
        diagonal[x][0]   = lowestScoreSum;
        horizontal[x][0] = x*gapPenalty + startupGapPenalty;
        vertical[x][0]   = lowestScoreSum;
    }
    for(int y=0; y < stringYlen+1; y++){
        diagonal[0][y]   = lowestScoreSum;
        horizontal[0][y] = lowestScoreSum;
        vertical[0][y]   = y*gapPenalty + startupGapPenalty;
    }
    diagonal[0][0] = 0;
    // Inductive step.
    double oneScore;
    for(int x=1; x < stringXlen+1; x++){
        for(int y=1; y < stringYlen+1; y++){
            if(stringX[x-1] == stringY[y-1])
                oneScore = matchScore;
            else oneScore = mismatchPenalty;
            diagonal[x][y] = oneScore
                + max( diagonal[x-1][y-1],
                       horizontal[x-1][y-1],
                       vertical[x-1][y-1]  );
            horizontal[x][y] = gapPenalty
                + max( diagonal[x-1][y]    + startupGapPenalty,
                       horizontal[x-1][y],
                       vertical[x-1][y]    + startupGapPenalty);
            vertical[x][y] = gapPenalty
                + max( diagonal[x][y-1]    + startupGapPenalty,
                       horizontal[x][y-1]  + startupGapPenalty,
                       vertical[x][y-1]);
        }
    }
}
```

Fig. 6.13 Gotoh's algorithm.

```
// Restore the alignment from the bottom-right to the top-left.
int x = stringXlen; int y = stringYlen;
char traceType;
double trace = max(diagonal[x][y], horizontal[x][y], vertical[x][y]);
// Move in the vertical or horizontal direction if there are multiple choices.
if(trace == vertical[x][y])              traceType = 'v';
else if(trace == horizontal[x][y])       traceType = 'h';
else                                     traceType = 'd';
int[][] alignment = new int[stringXlen+stringYlen][2];
int i = 0;
for( ; !(x == 0 && y == 0); i++){
    // No need to perform an array boundary check when x-1 and y-1 are accessed.
    switch(traceType){
    case 'd':
        if(stringX[x-1] == stringY[y-1])
            oneScore = matchScore;
        else oneScore = mismatchPenalty;
        if(diagonal[x][y] == oneScore + diagonal[x-1][y-1])
            traceType = 'd';
        else if(diagonal[x][y] == oneScore + horizontal[x-1][y-1])
            traceType = 'h';
        else traceType = 'v';
        alignment[i][0] = stringX[x-1]; alignment[i][1] = stringY[y-1];
        x = x-1; y = y-1;
        break;
    case 'h':
        if(horizontal[x][y] ==
                gapPenalty + diagonal[x-1][y] + startupGapPenalty)
            traceType = 'd';
        else if(horizontal[x][y] == gapPenalty + horizontal[x-1][y])
            traceType = 'h';
        else traceType = 'v';
        alignment[i][0] = stringX[x-1]; alignment[i][1] = -1; x = x-1;
        break;
    case 'v':
        if(vertical[x][y] ==
                gapPenalty + diagonal[x][y-1] + startupGapPenalty)
            traceType = 'd';
        else if(vertical[x][y] ==
                gapPenalty + horizontal[x][y-1] + startupGapPenalty)
            traceType = 'h';
        else traceType = 'v';
        alignment[i][0] = -1; alignment[i][1] = stringY[y-1]; y = y-1;
        break;
    }
}
printAlignment(alignment, i-1);
}
```

Fig. 6.14 Gotoh's algorithm.

back to the previous node and restoring the type of the last move among `diagonal`, `horizontal`, and `vertical` until it hits the top-leftmost node.

Note that Gotoh's algorithm computes elements in three matrices of size $(m+1)(n+1)$, where m and n are the lengths of the input two strings; then, it traces back the best alignment path by visiting at most $m+n+1$ nodes. Therefore, the worst-case computational complexity of Gotoh's algorithm is $O(mn)$.

6.8 Hirschberg's Space Reduction Technique

The dynamic programming algorithms that have been explained so far use matrices of size $(m+1)(n+1)$ for input strings of lengths m and n. Those matrices may be too large to fit into the main memory if the two input strings are very long. In 1975, Daniel S. Hirschberg presented an idea for reducing the space size at the expense of doubling the computational time [36]. Here, we illustrate Hirschberg's idea by modifying the Needleman-Wunsch algorithm.

The Needleman-Wunsch algorithm keeps all the elements in the score matrix for restoring the best global alignment path. The crux of Hirschberg's technique is to develop a space-efficient algorithm that does not attempt to compute the best global alignment path completely, but locates the node at which the best global alignment path goes through the middle column. Let us call the node in the middle column the *footprint* of the path.

The footprint is then used to divide the matrix into four submatrices. Note that the best global alignment path must go through the upper-left and lower-right matrices, as shown in the lower-right portion of Figure 6.15; therefore, the total size of these two matrices halves the size of the original matrix. Then, we can apply the algorithm to the individual submatrices to locate the footprints of the best alignment path, and iterate the step until locating the footprints in all columns, thereby restoring the best path. The overall computation time is proportional to the total size of all matrices scanned, which amounts to

$$(m+1)(n+1)(1 + \frac{1}{2} + \frac{1}{4} + \ldots) \leq 2(m+1)(n+1),$$

confirming that Hirschberg's method only doubles the computation time of the Needleman-Wunsch algorithm.

Here, we present a space-efficient algorithm that computes the footprint

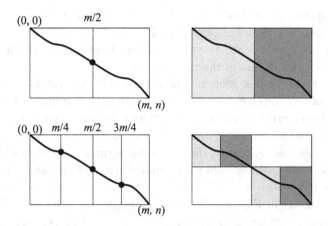

Fig. 6.15 The Hirschberg space-reduction technique uses less memory by memorizing only the footprint in the middle column of the focal matrix.

in the middle column in $O(mn)$ time using a matrix of size $O(\min(m,n))$, which is a dramatic space reduction. Let (x, y) denote the node at the x-th column and the y-th row. Let us consider the best global alignment path that starts at the top-leftmost node $(0, 0)$ and ends at the bottom-rightmost node (m, n). If the footprint in the middle column is $(m/2, k)$, where $m/2$ is the integer quotient, the path is illustrated by a chain of nodes:

$$(0,0) \to \ldots \to (m/2, k) \to \ldots \to (m, n)$$

If we scan the matrix in reverse order from the bottom-rightmost node, we will get the following path with the same best score because the score of a path is the linear sum of the scores of individual moves along the path.

$$(0,0) \leftarrow \ldots \leftarrow (m/2, k) \leftarrow \ldots \leftarrow (m, n)$$

Now, let us consider the following two paths that approach the footprint $(m/2, k)$ from both ends:

$$(0,0) \to \ldots \to (m/2, k) \leftarrow \ldots \leftarrow (m, n)$$

Summing the scores of all moves in these two paths also gives the same best score, which also maximizes the sum of all scores in the following two paths for $i = 0, \ldots, n$:

$$(0,0) \to \ldots \to (m/2, i) \leftarrow \ldots \leftarrow (m, n)$$

In this way, to find the footprint in the middle column, we can compute the scores of the best global alignment paths from $(0,0)$ to $(m/2, i)$ and from (m, n) to $(m/2, i)$ for each $i = 0, \ldots, n$, and detect the footprint $(m/2, i)$ that maximizes the sum of the two best scores.

To calculate the best score to a node instead of the best path, we do not have to store all the elements in the score matrix; it is sufficient to keep the elements in the previous and current columns, requiring only memory space for data in two columns. Without the loss of generality, we assume that $m \geq n$; otherwise, exchange the two input strings to meet the criterion. Then, the height of the column is $\min(m, n)$, and the algorithm needs $O(\min(m, n))$ space.

The program in Figures 6.16 and 6.17 implements this idea. The program assumes that stringX is longer than or equal to stringY, and that it utilizes the temporary score table holding the elements in three columns of length stringY plus one. The first and second columns are used to store scores in the previous and present columns, and the third column keeps the best scores of paths to nodes in the middle column. Indexes prev and current alternatively refer to the first and second columns. NWHrecursion sets footprints to the array of footprints and footprintScores to the list of best scores to these footprints for the input submatrix specified by the upper-left corner (topLeftX, topLeftY) and the lower-right corner (bottomRightX, bottomRightY).

Problems

Problem 6.1   Give examples of useless series of edit operations.

Problem 6.2   A *subsequence* of a string is an ordered sequence of characters that may not be consecutive in the string. Modify the Needleman-Wunsch algorithm to output the longest common subsequence shared by two input strings.

Problem 6.3   Complete the traceback procedure in the algorithm for finding the best overlap alignment in Figure 6.10.

Problem 6.4   Modify Gotoh's algorithm to incorporate Hirschberg's space-reduction technique.

```java
public static void NeedlemanWunschHirschberg(int[] stringX, int[] stringY){
    double[][] tmpScore = new double[3][stringY.length+1];
    int[] footprints = new int[stringX.length+1];
    double[] footprintScores = new double[stringX.length+1];
    footprints[0] = 0;
    footprintScores[0] = 0;
    footprints[stringX.length] = stringY.length;
    NWHrecursion(footprints, footprintScores,
            tmpScore, stringX, stringY, 0, 0, stringX.length, stringY.length);
    // Print the score of the optimal alignment.
    (omitted)
}
public static void NWHrecursion(
            int[] footprints, double[] footprintScores,
            double[][] tmpScore, int[] stringX, int[] stringY,
            int topLeftX, int topLeftY, int bottomRightX, int bottomRightY){
    int middleX = (topLeftX + bottomRightX)/2; // Probe the middle vertical line.
    if(middleX == topLeftX || middleX == bottomRightX) return;

    // Scan from the top-left. Initialize the score array.
    int prev = 0; int current = 1;
    for(int y = topLeftY; y <= bottomRightY; y++)
        tmpScore[current][y] = gapPenalty*(y-topLeftY);
    // Inductive step.
    double oneScore;
    for( int x = topLeftX+1; x <= middleX; x++ ){
        prev = (prev+1)%2; current = (current+1)%2; // Exchange prev and current.
        tmpScore[current][topLeftY] = gapPenalty*(x-topLeftX);
        for( int y = topLeftY+1; y <= bottomRightY; y++){
            if(stringX[x-1] == stringY[y-1])
                oneScore = matchScore;
            else oneScore = mismatchPenalty;
            tmpScore[current][y] =
                max( tmpScore[prev][y-1]   + oneScore,    // Diagonal.
                     tmpScore[prev][y]     + gapPenalty,  // Horizontal.
                     tmpScore[current][y-1]+ gapPenalty); // Vertical.
        }
    }
    // Set aside the scores of the middle vertical line.
    int firstHalf = 2;
    for(int y = topLeftY; y <= bottomRightY; y++)
        tmpScore[firstHalf][y] = tmpScore[current][y];
```

Fig. 6.16 Hirschberg's space reduction idea for the Needleman-Wunsch algorithm.

```
// Scan from the bottom-right. Initialize the score array.
prev = 0; current = 1;
for(int y = bottomRightY; topLeftY <= y; y--)
    tmpScore[current][y] = gapPenalty*(bottomRightY-y);
// Inductive step.
for( int x = bottomRightX-1; middleX <= x; x-- ){
    prev = (prev+1)%2; current = (current+1)%2;
            // Exchange prev and current.
    tmpScore[current][bottomRightY] = gapPenalty*(bottomRightX-x);
    for( int y = bottomRightY-1; topLeftY <= y; y--){
        // Indexes of stringX and stringY are changed slightly.
        if(stringX[x] == stringY[y])
            oneScore = matchScore;
        else  oneScore = mismatchPenalty;
        tmpScore[current][y] =
            max( tmpScore[prev][y+1]   + oneScore,    // Diagonal.
                 tmpScore[prev][y]     + gapPenalty,  // Horizontal.
                 tmpScore[current][y+1]+ gapPenalty); // Vertical.
    }
}
// Set the footprint to the row of the maximum sum in the middle column.
int y = topLeftY; int middleY = topLeftY;
double maxSum = tmpScore[firstHalf][y]+tmpScore[current][y];
for( y++; y <= bottomRightY; y++){
    double currentSum = tmpScore[firstHalf][y]+tmpScore[current][y];
    if(maxSum < currentSum){ maxSum = currentSum; middleY = y; }
}
footprints[middleX] = middleY;
footprintScores[middleX]
        = footprintScores[topLeftX]+tmpScore[firstHalf][middleY];
footprintScores[bottomRightX] = footprintScores[topLeftX]+maxSum;

// Recursive application of this procedure to the upper-left half and
// lower-right half blocks.
NWHrecursion(footprints, footprintScores, tmpScore,
      stringX, stringY, topLeftX, topLeftY, middleX, middleY);
NWHrecursion(footprints, footprintScores, tmpScore,
      stringX, stringY, middleX, middleY, bottomRightX, bottomRightY);
}
```

Fig. 6.17 Hirschberg's space reduction idea for the Needleman-Wunsch algorithm.

Chapter 7
Seeded Alignments

The alignment of the growing set of cDNA sequences and expressed sequence tags with newly sequenced genomes reveals complex structures that involve exons and introns, and identifies alternatively spliced transcripts. For these purposes, the traditional dynamic programming (DP) local alignment algorithms introduced in the previous chapter may output a highly accurate alignment between two sequences, but they are computationally costly and infeasible in practice.

The seeded alignment technique illustrated in Figure 7.1 is several orders of magnitude faster than traditional DP algorithms because seeded alignment algorithms preprocess the long sequences in a database and build a large lookup table of short sequences called "seeds" in the main memory for determining the positions of seeds in the long sequences efficiently. The query sequence is then scanned to associate individual seeds in the query with their positions in the database, thereby listing candidate regions where the query and database are likely similar. Finally, we attempt to compute precise alignments by extending each seed hit in both directions, as illustrated in the lower portion of Figure 7.1. We call this the *seed and extend strategy*.

Seeded alignment methods should be "sensitive," so that the correct answer appears in one of the candidate regions with a high probability, and "efficient" in reducing the number of candidate regions. To achieve better sensitivity and specificity, a wide variety of seed patterns have been devised for software programs.

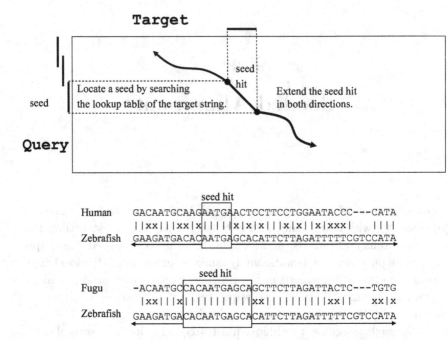

Fig. 7.1 The upper picture shows the basic concept of seeded alignments. The long target string is scanned to generate a lookup table of k-mer seeds. Subsequently, we search the query for an exact match for some k-mer seed and locate the seed hit in the target string. Using the approximate alignment algorithms, we extend the seed hit in both directions to detect a meaningful alignment. The lower picture shows two alignments of the human homeobox gene *HOXA1* with its homologous regions in the fugu and zebrafish shown in Figure 6.1. The seeded alignment technique finds the seeds enclosed in boxes and calculates their extensions.

7.1 Sensitivity and Specificity

We define the sensitivity of seeds to evaluate their quality. Here, we assume that a match of two letters at every position in the query and its homologous string is modeled as an independent, identically distributed Bernoulli random variable, and let M denote the match probability. To process genomic sequences that comprise fairly diverse strings, the Bernoulli random distribution is appropriate. When aligning cDNA sequences, however, the dependence of the bases within a codon must be considered, and more involved distributions, such as a Markov chain model [12], are more appropriate.

Definition 7.1 Let us compare a given query of length q and its homologous region of the same length. A comparison is made to check if the two letters at the same position match. Inserting gaps into the two strings is not considered. Let M denote the probability of two letters matching at every position. We define the *sensitivity* of seeds as the probability that the query and its homologous region of the query length share at least one seed in common. The sensitivity of each seed is determined from the value of the query length and match ratio.

In this definition, we disallow the insertion of gaps into the two strings. This is partly because mismatches are observed more frequently in genomes than insertions and deletions by one order of magnitude. Another reason is that the analysis of sensitivity is much simpler if only mismatches are processed.

The next question is how do we measure the specificity of a seed? If a region and the query share a seed, the region is a homologous candidate. However, such a candidate could be a false positive that yields no meaningful alignment because it happens to share the seed with the query. Therefore, a lower probability that the query and an arbitrary region of the same length happen to share a seed implies a higher specificity of using the seed, and we call the probability the *specificity* of the seed. To calculate the probability, we assume that letters are distributed uniformly in the database. When nucleotide sequences are processed, the probability equals the sensitivity when the match ratio M is set to $1/4$.

Definition 7.2 The *specificity* of finding a k-mer is the probability that at least one occurrence of the seed is found in the query when the letters are distributed uniformly and two letters match with a probability of $1/4$ when nucleotide sequences are processed.

Three parameters affect the sensitivity and specificity: the length of seeds, query length, and match ratio. The use of shorter seeds, longer queries, or higher match ratios increases the probability that the query and its homologous region share at least one seed, thereby increasing the sensitivity. However, the improvement is a double-edged sword because the false-positive rate also increases, reducing the specificity.

To understand the trade-off between sensitivity and specificity, Table 7.1 presents the sensitivity and specificity of perfectly matching k-mers for a variety of match ratios when the queries are 20, 50, and 100 in length. In the next section, algorithms for computing these values are presented. The motivation behind the selection of lengths 20, 50, and 100 is that

Table 7.1 The sensitivity and specificity of a perfect match of k-mers for a variety of match ratios ranging from 80 to 100% when the queries are 20, 50, and 100 in length.

Match Ratio	k-mer seeds									
	3	4	5	6	7	8	9	10	11	12
100%	1.000	1.000	1.000	1.000	1.000	1.000	1.000	1.000	1.000	1.000
98%	1.000	1.000	1.000	1.000	0.999	0.995	0.989	0.980	0.945	0.910
96%	1.000	1.000	1.000	0.998	0.993	0.979	0.958	0.931	0.868	0.809
94%	1.000	1.000	0.999	0.994	0.980	0.951	0.911	0.862	0.780	0.704
92%	1.000	1.000	0.997	0.986	0.959	0.912	0.851	0.782	0.687	0.603
90%	1.000	0.999	0.993	0.972	0.929	0.862	0.782	0.697	0.596	0.508
88%	1.000	0.998	0.987	0.952	0.890	0.804	0.709	0.613	0.510	0.423
86%	1.000	0.997	0.977	0.926	0.844	0.741	0.634	0.531	0.430	0.347
84%	1.000	0.993	0.962	0.893	0.792	0.675	0.560	0.455	0.358	0.281
82%	0.999	0.989	0.943	0.854	0.735	0.608	0.488	0.385	0.295	0.226
80%	0.999	0.982	0.920	0.810	0.676	0.541	0.422	0.322	0.241	0.179
Specificity	2.00E-01	5.01E-02	1.19E-02	2.81E-03	6.56E-04	1.53E-04	3.53E-05	8.11E-06	1.85E-06	4.17E-07

$Q = 20$

Match Ratio	k-mer seeds									
	6	7	8	9	10	11	12	13	14	15
100%	1.000	1.000	1.000	1.000	1.000	1.000	1.000	1.000	1.000	1.000
98%	1.000	1.000	1.000	1.000	1.000	1.000	1.000	1.000	0.999	0.998
96%	1.000	1.000	1.000	1.000	1.000	0.999	0.997	0.994	0.988	0.979
94%	1.000	1.000	1.000	0.999	0.998	0.994	0.986	0.974	0.957	0.933
92%	1.000	1.000	0.999	0.997	0.991	0.980	0.961	0.935	0.900	0.859
90%	1.000	1.000	0.997	0.991	0.977	0.953	0.918	0.873	0.820	0.762
88%	1.000	0.998	0.992	0.978	0.950	0.909	0.856	0.793	0.724	0.654
86%	0.999	0.995	0.983	0.955	0.911	0.850	0.778	0.700	0.621	0.544
84%	0.998	0.990	0.966	0.922	0.857	0.778	0.691	0.602	0.518	0.440
82%	0.996	0.979	0.940	0.876	0.792	0.696	0.599	0.506	0.421	0.347
80%	0.991	0.963	0.905	0.819	0.717	0.610	0.507	0.415	0.334	0.267
Specificity	8.28E-03	2.03E-03	4.96E-04	1.21E-04	2.96E-05	7.21E-06	1.76E-06	4.28E-07	1.04E-07	2.54E-08

$Q = 50$

Match Ratio	k-mer seeds									
	8	9	10	11	12	13	14	15	16	17
100%	1.000	1.000	1.000	1.000	1.000	1.000	1.000	1.000	1.000	1.000
98%	1.000	1.000	1.000	1.000	1.000	1.000	1.000	1.000	1.000	1.000
96%	1.000	1.000	1.000	1.000	1.000	1.000	1.000	1.000	1.000	0.999
94%	1.000	1.000	1.000	1.000	1.000	1.000	0.999	0.998	0.995	0.991
92%	1.000	1.000	1.000	1.000	0.999	0.998	0.994	0.988	0.977	0.963
90%	1.000	1.000	1.000	0.999	0.996	0.989	0.978	0.961	0.936	0.904
88%	1.000	1.000	0.998	0.994	0.986	0.969	0.944	0.909	0.865	0.814
86%	1.000	0.999	0.994	0.984	0.963	0.931	0.886	0.831	0.768	0.700
84%	0.999	0.996	0.985	0.962	0.924	0.871	0.805	0.731	0.654	0.577
82%	0.997	0.988	0.966	0.925	0.865	0.791	0.707	0.620	0.536	0.456
80%	0.993	0.974	0.934	0.871	0.789	0.696	0.600	0.508	0.423	0.348
Specificity	1.07E-03	2.64E-04	6.53E-05	1.62E-05	3.99E-06	9.87E-07	2.44E-07	6.03E-08	1.49E-08	3.68E-09

$Q = 100$

nucleotide sequences 20 bases in length are typically used as PCR primers for amplifying particular sequences. The sequences spotted on microarrays are about 50 bases long. The average length of exons in the human genome is about 150.

For example, let us enumerate all positions where the given 20-mer string may match with $\geq 80\%$ identity. According to Table 7.1, the use of 3-mer seeds allows us to identify positions with a probability of 0.999, but the sensitivity is 2.00×10^{-1}, implying that the 20-mer query and an arbitrary 20-mer share at least one 3-mer seed with a probability of 0.2, which is too high to narrow the set of candidates effectively. We can increase the specificity at the expense of the sensitivity by using longer seeds; e.g., if we use 10-mer seeds, the specificity becomes 8.11×10^{-6}, while the sensitivity falls to 0.322, which is very low.

If the query length is 100, the use of 10-mer seeds would be effective even if the match ratio is 80% because the sensitivity is 0.934, and the specificity is 6.53×10^{-5}. The choice of the match ratio is dependent on the problem that needs to be solved. The match ratio can be set to a value lower than the average match ratio so as to reveal meaningful similarity. To compare the mouse and rat genomes, one might choose an 80% match ratio, while to compare the human and chimpanzee genomes, the match ratio could be set to, say, 94%. In this case, if the query is 100 in length, 17-mer seeds would be sufficient because the sensitivity is 0.991 and the specificity is 3.68×10^{-9}.

The meaning of the specificity may not be straightforward. A specificity of 3.68×10^{-9} means that the 100-mer query matches an arbitrary 100-mer with a probability of 3.68×10^{-9}. If a target genome of length 3×10^9 is studied, there are about 3×10^9 100-mer strings in the genome. Therefore, the estimated number of arbitrary 100-mers that the 100-mer query may share with a 17-mer seed is $3 \times 10^9 \times 3.68 \times 10^{-9} = 11.04$. In this way, the product of the specificity and the total length of the target sequence provides us with the estimated number of candidate positions where the query might match with the given match ratio.

7.2 Computing Sensitivity and Specificity

In what follows, we present an $O(q^4)$-time dynamic programming algorithm for computing the sensitivity of finding at least one k-mer seed, which was designed by Kasahara (see [72]). Let T be a three-dimensional array and let

$T[q][m][k]$ denote the number of strings of length q, such that each sequence contains m mismatches and the longest perfect matches between the query and its homologous region are k-mers. $T[q][m][k]$ is then defined inductively as follows.

The base case is when $m = 0$, which indicates there are no mismatches; this does not make sense unless $q = k$. In particular, when $q = 0$, the query is the empty string, and it appears useless to define a value for this case, but we assume that $T[q][0][k]$ is defined for $q = 0$, which will be explained shortly, according to the following formula:

$$T[q][0][k] = \begin{cases} 1 \text{ if } q = k \\ 0 \text{ otherwise} \end{cases}$$

In the inductive step, we assume $m \geq 1$, and we divide the problem into two cases: when a k-mer matches at the rightmost position and otherwise (see Figure 7.2).

$$T[q][m][k] = \sum_{j=0}^{\min(k,q-m-k)} T[q-k-1][m-1][j] + \sum_{i=0}^{\min(k-1,q-m-k)} T[q-i-1][m-1][k]$$

If a k-mer matches at the rightmost position, we move on to consider the left string of length $q - k - 1$ to allow the length of the longest perfect matches, j, to range from 0 to k. In contrast, if an i-mer ($i < k$) matches at the rightmost position, the left string is of length $q - i - 1$ and it must contain a k-mer match. It is essential that $T[q][0][k]$ is defined for $q = 0$. Consider the case when the last mismatch is located at the leftmost position and $j = 0$. To count this case, we need to set $T[0][0][0] = 1$. This inductive step dominates the overall calculation. For all values between 0 and q in each of the three arguments, $T[q][m][k]$ must be calculated. The total computation time is $O(q^4)$ because an individual value is computed in $O(q)$ time.

The sensitivity (probability) that at least one k-mer is shared by the query of length q and its homologous region for the match ratio M is:

$$\sum_{m=0}^{q} \sum_{i=k}^{q} T[q][m][i] M^{q-m}(1-M)^m$$

Recall that we define the specificity of finding a k-mer as the probability that at least one occurrence of the seed is found in the query when the

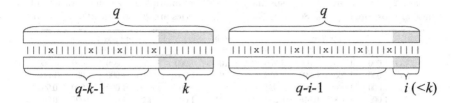

Fig. 7.2 The left picture illustrates the case when a k-mer matches at the rightmost position, and the total number in this case equals $\sum_{j=0}^{\min(k,q-m-k)} T[q-k-1][m-1][j]$. The right picture illustrates the other case, when the number of consecutive matches at the rightmost position is i, which is less than k, and the total number in this case amounts to $\sum_{i=0}^{\min(k-1,q-m-k)} T[q-i-1][m-1][k]$.

letters are distributed uniformly and two letters match with a probability of 1/4. Therefore, the specificity is obtained as the sensitivity by setting M to 1/4. Although assuming a uniform distribution does not hold in reality due to the many repeated sequences in a genome, the specificity criterion provides a relatively fair standard for comparing the specificities of various seeds.

Many widely used tools implement the idea of searching for a single hit of seeds between the query and its homologous region, such as BLAST [2], sim4 [30], SSAHA [71], and MegaBlast [107]. Seed length can also be variable; e.g., MUMmer 2 [26] and QUASAR [13] use suffix trees or suffix arrays to search for consecutive strings of variable length.

7.3 Multiple Hits of Seeds

More involved criteria require multiple seed matches, as in GappedBlast [3] and BLAT [47]. This approach compares every k-mer in the query with all k-mers in the target string using lookup tables. It seeks multiple perfect matches that are in close proximity and attempts to identify homologous regions that share multiple k-mers with the query.

The sensitivity and specificity of seeking multiple k-mers are defined similarly to the terms for finding a single k-mer; i.e., the sensitivity is the probability that the query and its homologous region share common multiple k-mers that do not overlap, and the specificity is the sensitivity when the match ratio of two letters at every position is 1/4. It is important to assume that the multiple k-mers are disjoint; otherwise, one $k+1$-mer might yield two k-mers that share $k-1$ consecutive letters.

Table 7.2 The sensitivity and specificity of seeking multiple k-mers for a variety of match ratios ranging from 80 to 100% when the queries are 20, 50, and 100 in length.

Match Ratio	\multicolumn{4}{c}{k-mer seeds}			
	3	4	5	6
100%	1.000	1.000	1.000	1.000
98%	1.000	1.000	0.999	0.988
96%	1.000	1.000	0.992	0.952
94%	1.000	0.998	0.976	0.897
92%	1.000	0.994	0.949	0.827
90%	1.000	0.986	0.911	0.747
88%	0.999	0.972	0.861	0.663
86%	0.997	0.953	0.803	0.578
84%	0.995	0.927	0.738	0.496
82%	0.990	0.893	0.669	0.419
80%	0.984	0.852	0.598	0.348
Specificity	1.68E-02	8.49E-04	3.94E-05	1.71E-06

$Q = 20$

Match Ratio	k-mer seeds				
	5	6	7	8	9
100%	1.000	1.000	1.000	1.000	1.000
98%	1.000	1.000	1.000	1.000	1.000
96%	1.000	1.000	1.000	1.000	0.998
94%	1.000	1.000	1.000	0.997	0.987
92%	1.000	1.000	0.998	0.987	0.960
90%	1.000	0.999	0.992	0.966	0.910
88%	1.000	0.996	0.978	0.927	0.837
86%	0.999	0.990	0.953	0.870	0.745
84%	0.998	0.978	0.914	0.794	0.640
82%	0.994	0.958	0.859	0.705	0.532
80%	0.988	0.927	0.790	0.608	0.428
Specificity	4.70E-04	2.69E-05	1.52E-06	8.56E-08	4.77E-09

$Q = 50$

Match Ratio	k-mer seeds									
	5	6	7	8	9	10	11	12	13	14
100%	1.000	1.000	1.000	1.000	1.000	1.000	1.000	1.000	1.000	1.000
98%	1.000	1.000	1.000	1.000	1.000	1.000	1.000	1.000	1.000	1.000
96%	1.000	1.000	1.000	1.000	1.000	1.000	1.000	1.000	1.000	0.998
94%	1.000	1.000	1.000	1.000	1.000	1.000	0.999	0.998	0.994	0.985
92%	1.000	1.000	1.000	1.000	1.000	0.998	0.994	0.986	0.968	0.938
90%	1.000	1.000	1.000	0.999	0.997	0.991	0.976	0.949	0.905	0.844
88%	1.000	1.000	0.999	0.997	0.989	0.970	0.934	0.877	0.799	0.706
86%	1.000	1.000	0.997	0.989	0.969	0.927	0.859	0.768	0.661	0.550
84%	1.000	0.998	0.992	0.973	0.930	0.857	0.754	0.635	0.513	0.399
82%	1.000	0.996	0.981	0.943	0.869	0.761	0.630	0.496	0.374	0.272
80%	0.999	0.990	0.961	0.895	0.786	0.646	0.500	0.366	0.257	0.175
Specificity	1.66E-03	1.03E-04	6.28E-06	3.81E-07	2.30E-08	1.39E-09	8.41E-11	5.07E-12	3.05E-13	1.83E-14

$Q = 100$

Table 7.3 The comparison between a single hit and multiple hits.

	type of seeds	sensitivity	specificity
$Q = 20, M = 80\%$	4-mer single hit	0.982	5.01E-02
	3-mer multiple hits	0.984	1.68E-02
$Q = 50, M = 80\%$	7-mer single hit	0.963	2.03E-03
	5-mer multiple hits	0.988	4.70E-04
$Q = 100, M = 80\%$	9-mer single hit	0.974	2.64E-04
	6-mer multiple hits	0.990	1.03E-04

Table 7.2 shows the sensitivity and specificity of finding multiple k-mers. A comparison to seeking a single seed hit in Table 7.1 shows that a single hit of a k_1-mer seed is often outperformed by multiple hits of k_2-mer seeds

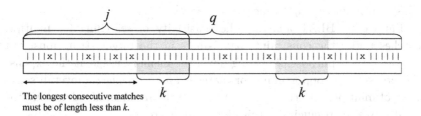

Fig. 7.3 We solve the problem of computing the sensitivity of multiple hits of seeds by dividing the query into two substrings according to the position of the first k-mer match.

in terms of both sensitivity and specificity, as listed in Table 7.3.

To compute the sensitivity of multiple hits of seeds, we divide all patterns of multiple k-mer hits into disjoint subsets according to the total number of mismatches, m, as the primary condition and the position of the first k-mer match from the leftmost as the secondary condition [27], as illustrated in Figure 7.3.

Suppose that the first k-mer match occurs at the right end of the first j-long substring of the query, implying that all other consecutive matches in the first j-long substring are of length less than k. Suppose that the first j-long substring contains $m_1 (0 \leq m_1 \leq j)$ mismatches, and let $First(j, m_1, k)$ denote the total number of such patterns of the first j-long substring, which is computed using T as follows but is not stored in an array.

$$First(j, 0, k) = \begin{cases} 1 \text{ if } j = k \\ 0 \text{ otherwise} \end{cases}$$

$$First(j, m_1, k) = \sum_{i=0}^{\min(k-1, j-m_1-k)} T[j-k-1][m_1-1][i] \quad \text{if } m_1 \geq 1$$

The rest $(q-j)$-long substring must contain $m - m_1$ mismatches and at least one k-mer match, and the total number of such patterns in the rest is

$$\sum_{i=k}^{q-j-(m-m_1)} T[q-j][m-m_1][i]$$

Finally, the sensitivity of finding multiple k-mers between a query of length q and its homologous region for match ratio M is

$$\sum_{m=0}^{q} \sum_{j=k}^{q-k} \sum_{m_1=0}^{\min(m,j)} First(j, m_1, k) \cdot \left(\sum_{i=k}^{q-j-(m-m_1)} T[q-j][m-m_1][i] \right) \cdot M^{q-m}(1-M)^m$$

This step requires $O(q^4)$ time.

Jim Kent's BLAT analyzes the sensitivity of searching for multiple matches under the constraint that k-mers in the target string must not overlap, for the purpose of cutting down the memory usage at the expense of decreasing the opportunity for matches and reducing the overall sensitivity of multiple-hit criteria [47]. We leave the problem of calculating the sensitivity and specificity of this case as an exercise.

7.4 Gapped Seeds

Another promising class of seeds is spaced seeds (nonconsecutive strings), which allow some fixed-length spaces for mismatches and require matching pairs at only a subset of all positions. This concept is used to solve the approximate string matching in the k-mismatches problem that finds all strings matching the query sequence with at most k mismatches [14, 16, 77]. Recently, Ma, Tromp, and Li utilized spaced seeds to improve the sensitivity of seeded alignments [58], and BLASTZ used spaced seeds to compare the human and mouse genomes [90]. For example, seed pattern 11*111**11 requires matches at positions 0, 1, 3, 4, 5, 8 and 9, where * indicates "don't care" positions that do not have to match. The following alignment shows two strings that share 11*111**11:

```
AACTTTAACC
|| |||  ||
AAGTTTGGCC
```

Optimally spaced seeds that maximize the sensitivity of detecting single spaced seed matches in homologous regions are the most valuable [58], and this has led many researchers to develop exponential or heuristic algorithms for computing optimal ones [12, 18, 19, 55]. For example, Buhler, Keich, and Sun developed a general framework for evaluating the sensitivity of spaced seed criteria in a k-th order Markov model [12]. Choi, Zeng, and Zhang presented optimally spaced seeds for several seed lengths and for various match ratios between homologous regions [18]. In general, both calculating optimally spaced seeds that maximize the sensitivity of finding single matches and computing the sensitivity of a particular spaced seed are intractable to compute [55]. For more details, see a tutorial by Brown, Li, and Ma [11].

7.5 Chaining and Comparative Genomics

In Section 6.6, we explained the problem of aligning cDNA sequences with a genome. Gotoh's dynamic programming algorithm in Section 6.7 is one solution to this problem, but it is not feasible for handling a large genome because the algorithm runs in $O(mn)$ time, where m is the cDNA length and n is the genome length. We need an algorithm that works in time almost proportional to the genome size, n, and outputs a nearly optimal solution at the expense of seeking the optimal answer.

Seeds are useful in designing such an efficient algorithm. The key concept is that we locate all the seeds of the query cDNA sequence in the genome by consulting the lookup table of seeds, and subsequently attempt to *chain* a subset of these seed hits to calculate a nearly optimal alignment. Figure 7.4 illustrates this idea. In the figure, seed hits are shown as dots or lines that indicate long consecutive hits. Seed matches are not necessarily contiguous because of mismatches, insertions, and deletions, but the result gives a global picture of local alignments between the two sequences without using a costly dynamic programming approach. Although the genome is extremely large, if longer seeds are used, the number of seed hits is likely to be small enough to make the chaining process executable in practice. Of course, care must be taken to select moderate seed lengths to avoid overlooking meaningful matches.

Another important application of chaining seed hits is comparative genomics. This is an extremely computationally intensive task because genomes are much longer than transcripts, e.g., the size of the human

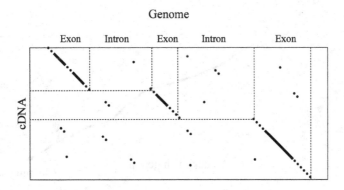

Fig. 7.4 An edit graph for the alignment between a cDNA sequence and a genome.

genome is about 3×10^9 base pairs, while human transcripts have at most 10^5 bases. In Figure 7.5, each dot represents a local alignment between parts of the medaka fish and pufferfish genomes. These local alignments are obtained by mapping nonoverlapping 300-base-long substrings in the pufferfish genome into the medaka genome using the seeded alignment technique. Alignments with $\geq 60\%$ identity are displayed. The curved arrow indicates one long chain of homologous regions.

The length of queries and the threshold of the identity ratio must be selected carefully by considering the differences between the two genomes. For example, if we compare the genomes of the human and chimpanzee, which diverged from a common ancestor about five million years ago, the average match ratio between these two genomes is $> 97\%$ [91], and hence we can safely use such longer seeds for which both the sensitivity and specificity are sufficiently high. Conversely, if we compare the genomes of the human and zebrafish, which diverged from a common ancestor about 450 million years ago, so as not to overlook low but significant similarity, we need to use shorter seeds at the sacrifice of computational performance.

Next, we introduce some terms related to chains.

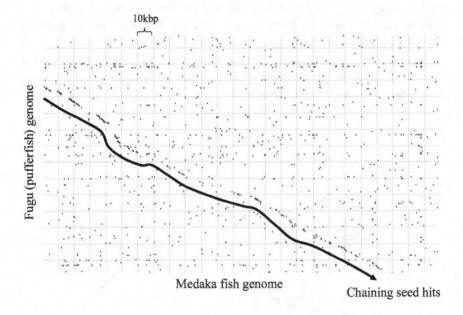

Fig. 7.5 An edit graph that compares parts of the medaka and pufferfish genomes.

Definition 7.3 Let T and Q be two input strings. A local alignment of the substrings $T[s_x, e_x]$ and $Q[s_y, e_y]$ is represented as a path from (s_x, s_y) to (e_x, e_y) in the edit graph, and let $(s_x, s_y) \rightarrow (e_x, e_y)$ denote the local alignment. (s_x, s_y) and (e_x, e_y) are called the *startpoint* and *endpoint* of A. Local alignment A, $(s_x, s_y) \rightarrow (e_x, e_y)$, is a *predecessor* of local alignment A', $(s'_x, s'_y) \rightarrow (e'_x, e'_y)$, if both $e_x \leq s'_x$ and $e_y \leq s'_y$ hold. This predecessor relationship is denoted by $A \prec A'$, and A' is called a *successor* of A. A *chain* is a sequence of alignments $A_1 \prec A_2 \prec \ldots \prec A_k$.

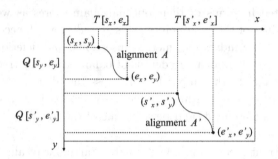

Fig. 7.6 A predecessor relationship $A \succ A'$ between two local alignments.

In this definition, seed hits can be treated as alignments at 100% identity. Given a set of alignments, chaining is the process of finding good chains in terms of some criteria. When scores are associated with alignments, a simple measure for evaluating a chain is the sum of the scores in the alignments of the chain [41]. To design more elaborate measures, one can consider gap scores between two alignments in the chain [28, 68] or allow some overlap between alignments in the chain [95]. Another direction of extension is to handle more than two input sequences [68]. Abouelhoda and Ohlebusch described a comprehensive survey of chaining algorithms [1].

Here, we use the simple measure of taking the sum of the scores. The upper picture in Figure 7.7 illustrates alignments labeled with scores. Directed arcs are drawn from each alignment to its possible candidates of predecessor alignments. The score of the optimal chain is 35, which is the sum of 10, 11, and 14. Here, we present an efficient algorithm for computing the optimal chain in terms of the simple measure in $O(k \log k)$ time, where k is the number of given alignments. The lower picture in Figure 7.7 illustrates the operation of the algorithm on the upper example of alignments.

Algorithm 7.1 Given a set of k alignments, we assign to each alignment A the score of the optimal chain among all chains that end with A, which we call the *optimal chain score* of A. For each alignment in the form $(s_x, s_y) \rightarrow (e_x, e_y)$, put x-coordinate values s_x and e_x into a set X and sort X in ascending order so that lower elements in X are processed first.

During the execution, we use another list, Y, that stores the sorted list of a subset of alignments according to the y-coordinate values of their endpoints. Y is used to select the predecessor of each alignment A that yields the optimal chain ending with A. Y is maintained so that alignments in Y are sorted in terms of their optimal chain scores as well as the y-coordinate values of their endpoints. Y is stored in a balanced binary tree in order to process individual queries, insertions, and deletions into Y in $O(\log k)$ time (see [23] for more details of balanced binary trees).

Initially, set Y to the empty list. Process the alignments from left to right by scanning the x-coordinate values in X. For each focal x-coordinate, let A denote its alignment, and perform one of the following two steps:

- If the current x-coordinate in X is the startpoint of alignment A, we calculate the optimal chain score of A. Let s_y denote the y-coordinate of the startpoint of A. If Y is not empty, conduct binary search Y for the closest alignment whose y-coordinate value is less than or equal to s_y. Since its optimal chain score is the maximum among possible predecessors, let the alignment be the predecessor of A and add its optimal chain score to the score of A.

  For example, alignment c selects a as its predecessor since Y has only a at the time of selection. The alignment f has two predecessor candidates in Y: a and c. Then, c is selected because c is closer to f and its score is greater than that of a.

- Otherwise, the current x-coordinate in X is the endpoint of A. Next, we consider if it is worthwhile to insert A into Y. Let e_y denote the y-coordinate of the endpoint of A. Search Y to check if there is a closest alignment such that the y-coordinate of its endpoint, e'_y, is less than or equal to e_y, namely $e'_y \leq e_y$. If there are no such alignments, insert A into Y to keep Y sorted. Otherwise, if the optimal chain score of A is greater than that of the alignment ending at e'_y, insert A into Y, while keeping Y sorted.

  Here, we show some alignments that are not inserted into Y. In Figure 7.7, the y-coordinate values of the endpoints of d and h are equal, and the optimal chain score of d is greater than that of h, indicating

Seeded Alignments

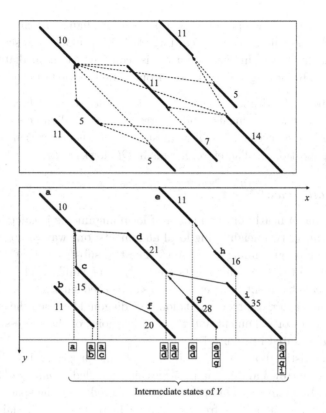

Fig. 7.7 The upper picture shows examples of alignments in an edit graph. The lower part illustrates the operation of the chaining algorithm on the upper example. As x-coordinate values are scanned from the left, the content of Y may change when the end of each alignment is processed. Dotted lines indicate the correspondence between the endpoints and intermediate states of Y. Numbers attached to alignments express scores in the upper picture and optimal chain scores in the lower portion.

that any successor of h is able to select d in place of h to increase its optimal chain score. Therefore, it is pointless to insert h into Y. Similarly, f is not inserted into Y in the presence of d.

Finally, if A is inserted into Y, scan Y to delete any alignment such that the y-coordinate of its endpoint $e_y^\#$ is greater than or equal to e_y ($e_y \leq e_y^\#$), and the optimal chain score of the alignment is less than that of A. For example, c is eliminated from Y when d is inserted because any successor of both c and d will select d in order to increase its score. Similarly, b and a are removed respectively when c and e are inserted. This operation guarantees that alignments in Y are also sorted in ascending order according to their optimal chain scores, and alignments to be eliminated must exist contiguously immediately after A in Y. Removing one alignment from Y takes $O(\log k)$ or $O(l)$ time.

The above algorithm searches Y for the predecessors of k alignments, inserts at most k alignments, and deletes at most k alignments, because each alignment is inserted into Y at most once, and hence, it is also deleted from Y at most once. Therefore, it runs in $O(k \log k)$ time.

Banded alignment

Provided that a nearly optimal chain of local alignments is calculated, we need to connect two neighboring local alignments; one way to do this is to use dynamic programming algorithms. The two substrings that are sandwiched by the two local alignments are likely to be similar. Therefore, the optimal local alignment between the two sandwiched substrings is expected to go through a fairly narrow band along the diagonal between the endpoint of the predecessor alignment and the starting point of the successor alignment, as illustrated in Figure 7.8.

This heuristic allows us to apply the dynamic programming algorithm to the gray, narrow band in Figure 7.8, which is called a *banded alignment*. The banded alignment technique is effective at reducing the space for edit graph construction, and is frequently used in practice. It is fairly simple to modify the Needleman-Wunsch algorithm to incorporate the banded alignment idea, and the problem is left to the reader as an exercise.

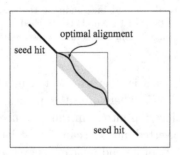

Fig. 7.8 The two bold lines show local alignments of seed hits. The rectangle indicates the edit graph used by dynamic programming algorithms, such as the Needleman-Wunsch algorithm. Considering the similarity between the two strings in this region, the optimal alignment goes through the narrow band along the diagonal.

7.6 Design of Highly Specific Oligomers

In what follows, we focus on an important application of seeded alignments. Sequence specificity is critical in molecular biological experiments. For example, to amplify a sequence using PCR, the PCR primer set that magnifies the target gene must be devised to hybridize the intended gene only and must not cross-react with any other gene. In analyzing the gene expression levels using a DNA microarray, the sequences of probes spotted on the microarray slide must hybridize the target sequence only, and to perform gene knockdown using RNAi, the double-stranded RNA must not silence an off-target gene. Therefore, it is essential to achieve sequence specificity, or to design the appropriate specific sequence or sets, to meet these various demands.

To design effective sequence tags, we must select sequences that hybridize a target sequence only and do not hybridize nontarget sequences. This is a prerequisite for designing primers, microarray oligo sequences, and siRNA sequences. Several methods of measuring specificity and selecting the appropriate hybridization model have been proposed.

Occurrence frequency and longest common factor

One simple measure is the occurrence frequency of a k-mer, which is the number of occurrences of the k-mer in the target sequence, and k-mers of occurrence frequency 1 appear only once and are treated as specific. In Section 4.5, we showed how to calculate the occurrence frequencies of all k-mers using the suffix array and the longest common prefix information, which runs in time proportional to the target sequence length.

For example, Figure 7.9 shows the distribution of k-mer occurrence frequencies in the human genome (build #30) [105]. In the graph, the x-axis shows the frequency F in log scale, while the y-axis presents the number of k-mers of frequency F in log scale. Each line denotes the number of k-mers of frequency F for $k = 8, 9, \ldots, 18$. Our primary interest is to enumerate unique k-mers that occur only once in the genome, and the number of unique k-mers is illustrated by the sequence of y-intercepts. For example, there are no unique 8-mers in the human genome; the least-frequent 8-mer appears about 330 times. Unique k-mers are observed when k is greater than 10. There are about 1,750 unique 11-mers, although there are too few to cover the entire genome. The number of unique k-mers increases with k, and there are about 1.66 billion unique 18-mers, which cover more than half

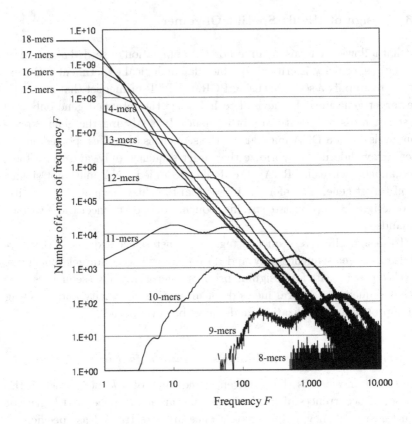

Fig. 7.9 Each line in the graph shows the number of k-mers (the vertical axis) of frequency F (the horizontal axis) for $k = 8, 9, \ldots, 18$ in the human genome (build #30) (*Journal of Bioinformatics and Computational Biology*, 2004, 2(1):36).

of the entire genome. The graph in Figure 7.10 indicates that the number starts to converge when k exceeds 18. Due to repetitive sequences in the human genome, the graphs in Figures 7.9 and 7.10 differ from those based on random nucleotide occurrence. Computing the random distribution is left as an exercise.

Although unique k-mers that appear only once in the genome is one measure of evaluating the specificity of k-mers, a unique k-mer could be highly similar to another string in the sense that the two strings share a long stretch of matching characters called a common factor in Section 4.5. The smaller length of the longest common factor given in Definition 4.5 indicates the higher specificity of the k-mer. Rahmann proposed using this

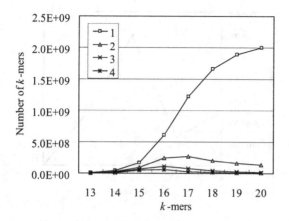

Fig. 7.10 Each line shows the number of k-mers ($k = 13, 14, \ldots, 20$) of frequency $F = 1, 2, 3, 4$.

measure to design microarray oligonucleotide probes [81].

Mismatch tolerance

The query string, such as a primer, oligomer, or siRNA sequence, may partially match the target string, i.e., all but a few nucleotides in the query are identical to the nucleotides in the target string. To find candidates for such off-target sequences, we introduce mismatch tolerance as a specificity measure. The mismatch tolerance is the minimum number of mismatches that allows the query to match the most similar off-target candidate sequence.

Definition 7.4 Given two strings, Q and Q', of the same length n, the Hamming distance $d_H(Q, Q')$ is defined as the number of positions at which the two bases differ, i.e.,

$$d_H(Q, Q') = |\{i \mid Q[i] \neq Q'[i], i = 1, \ldots, n\}|$$

Let T denote the target input string of length $|T|$, and let Q denote the query string of length $|Q|$ that is a substring of T. We define the *mismatch tolerance* of Q with respect to T, which is denoted by $mt(Q, T)$. If the occurrence frequency of Q in T exceeds 1, the unique location of Q in T cannot be specified; therefore, we set $mt(Q, T) = 0$. Otherwise, $mt(Q, T)$ is the minimum Hamming distance between Q and any subsequence of length $|Q|$ in T that is not identical to Q, i.e.,

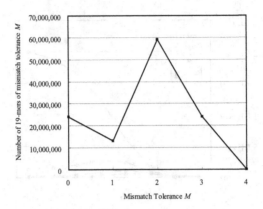

Fig. 7.11 The distribution of 19-mers with mismatch tolerance M on the horizontal axis (*Nucleic Acids Research*, 2004, 32(Web Server issue):W127).

$$mt(Q,T) = \min\{d_H(Q, T[i, i+|Q|-1]) \mid \\ i = 1, \ldots, |T|-|Q|+1, Q \neq T[i, i+|Q|-1]\}$$

Note that Q should not be compared with itself in T.

Example 7.1 Let T denote

ATGCTACGATCGTAGCTC,

and let Q be the first four letters in T, ATGC. Observe that $mt(Q,T) = 2$.

As a practical example, Figure 7.11 presents the distribution of 19-mers of mismatch tolerance in the human genes registered in the RefSeq and Unique UniGene databases [69]. The total length of these human genes approximates 1.2×10^8. The vertical axis shows the number of 19-mers of mismatch tolerance M on the horizontal axis. There are an ample numbers of 19-mers with mismatch tolerances of three or four, but there are no 19-mers with a mismatch tolerance of five or more. If we assume that nucleotides occur at random, the probability of observing 19-mers with a mismatch tolerance of zero is close to zero. However, such 19-mers are abundant because of repetitive or redundant sequences in the sequence database. Computing the random distribution is left as an exercise.

7.7 Seeds for Computing Mismatch Tolerance

Calculating the mismatch tolerance is computationally costly if we scan the entire target sequence. By using the lookup table of seeds in a large-scale target sequence, we can expedite the calculation of the mismatch tolerance.

(1) Lookup table creation phase: Build a lookup table that stores the locations of all the short seed sequences in T. This step is performed only once for T.
(2) Search phase: Search the lookup table for the positions of Q's seed sequences in T. The search trial of one seed effectively eliminates positions where Q cannot hybridize with m or less mismatches, and the search reports all the other potential locations to check. The seed length must be defined properly in each algorithm to avoid overlooking potential off-target sequences.
(3) Check phase: For individual seeds that hit T, which we call *seed hits*, try to align Q with T and count the number of mismatches in the alignment.

The overall performance is dominated by two factors: the total number of *seed search trials* in Step 2 and the total number of seed hits in Step 3. In general, there is a trade-off between these two major factors, i.e., an increase in one factor often decreases the other. Starting with a naive algorithm, we will now improve a suite of algorithms step-by-step.

7.7.1 Naive Algorithm

The idea of the naive algorithm is to look for a contiguous common string shared between the query string and its potential off-target sequences. The major issue is the choice of the minimum length of contiguous strings to search. For example, Figure 7.12 illustrates two alignments with four mismatches. In the left alignment, the length of the contiguous string shared between the two strings is eight, while the value is three in the right alignment.

In general, in order to return any off-target sequences Q' such that the Hamming distance $d_H(Q, Q') \leq m$, we use the property that Q and Q' must share a contiguous string of length $\left\lceil \frac{|Q|-m}{m+1} \right\rceil$, where $\lceil x \rceil$ denotes the smallest integer that is greater than or equal to x.

Example 7.2 When Q is a 19-mer string and $m = 4$, as illustrated in

```
TCAGTAGTATGTACTCCTA      TAGGCTGTGTATGATGCTC      query
|XX|||||||||X|||||X|     |||X|||X|||X|||X|||
TGCGTAGTATGCACTCCAA      TAGCCTGAGTAGGATACTC      off-target
```

Fig. 7.12 Alignments between a 19-mer query and potential off-target sequences with four mismatches. Boxes indicate the longest common matches.

the right alignment of Figure 7.12, the seed length is three.

In this section, we consider 19-mer queries because designing highly specific 19-mers is of practical use in designing siRNA sequences that match the target mRNA sequence to silence its behavior. Another reason is that typical primers and oligomers are of lengths greater than 19 and it is less difficult to guarantee that they are highly specific.

Let l denote the seed length $\left\lceil \frac{|Q|-m}{m+1} \right\rceil$. The number of seed search trials is $|Q| + 1 - l$. In order to estimate of the number of seed hits roughly, let us assume that all seeds are distributed uniformly in T. The number of positions where one seed occurs in T should be $|T|/4^l$. Multiplying this figure by the number of seed search trials gives the expected number of seed hits, $\frac{|T|}{4^l} \cdot (|Q| + 1 - l)$.

Example 7.3 When $|T| = 1.5 \times 10^8$, $|Q| = 19$, and $m = 4$, the seed length is three, the number of seed search trials is 17, and the expected number of seed hits is 39,843,750.

In this example, T is set to 1.5×10^8 because this would be an upper bound of the total length of all human mRNA sequences, according to the current knowledge.

7.7.2 BYP Method

The naive algorithm searches all $|Q|+1-l$ seeds of length l. Here, we present a concept for reducing the number of seed search trials, as is illustrated in Figure 7.13. In the figure, we see that if there are four mismatches, at least one of the five seeds enclosed in boxes must contain no mismatches.

Baeza-Yates and Perleberg generalized this idea and presented an algorithm called the BYP method that uses the following property [6].

Proposition 7.1 Let Q and Q' be sequences of the same length, such that the Hamming distance between Q and Q' is m. Q can be divided into

Fig. 7.13   Key concept of the BYP method.

disjoint substrings such that $(m+1)$ substrings are of length $\left\lfloor \frac{|Q|}{m+1} \right\rfloor$, where $\lfloor x \rfloor$ is the greatest integer that is less than or equal to x. At least one of them will exactly match Q' with no mismatches.

Example 7.4   For example, when $|Q| = |Q'| = 19$ and $m = 4$, five disjoint substrings of length three can be extracted, and at least one of them must match Q' exactly, as shown in Figure 7.13.

According to this proposition, using seeds of length $\left\lfloor \frac{|Q|}{m+1} \right\rfloor$, which is denoted by l in this subsection, allows us to find a seed that fully matches a potential off-target sequence, such as Q' in the proposition. Note that $\left\lfloor \frac{|Q|}{m+1} \right\rfloor$ equals $\left\lceil \frac{|Q|-m}{m+1} \right\rceil$, the seed length of the naive algorithm in the previous subsection. The number of seed search trials is $(m+1)$, and the expected number of seed hits under the assumption of a uniform seed distribution is $\frac{|T|}{4^l} \cdot (m+1)$.

Example 7.5   When $|T| = 1.5 \times 10^8$, $|Q| = 19$, and $m = 4$, the number of seed search trials is 5, and the expected number of seed hits is 11,718,750.

7.8   Partially Matching Seeds

The algorithms presented in the previous subsections require that at least one seed in Q exactly matches the off-target sequence Q'. As we have seen, the seeds are likely to be short, which makes it difficult to reduce the expected number of seed hits. Note that if we allow seeds to partially match off-target sequences with at most $s(\leq m)$ mismatches, we can use longer seeds. Sung and Lee used this property to select probes from large genomes [98]. First, we will introduce their idea, and then improve on it to achieve an acceleration of an order of magnitude.

For example, Figure 7.14 presents two alignments. In each alignment, one of the seeds of length nine contains at most two mismatches. The following proposition generalizes this property.

```
TGCGCAGCA TGTACTGCAC          GCCGTAGTA TAGGCTGCTC          query
| | | X | | X | | | | X | | X | | |   X | | X | | | X | | | | | | | X | | |
TGCGTAGTA TGTGCTCCAC          TCCCTAGAA TAGGCTACTC          off-target
```

Fig. 7.14 Partially matching seeds.

Proposition 7.2 If Q is divided into $\left\lceil \frac{m+1}{s+1} \right\rceil$ disjoint seeds of length $\left\lfloor \frac{|Q|}{\lceil \frac{m+1}{s+1} \rceil} \right\rfloor$, one of these substrings must match Q' with at most s mismatches.

Proof. Otherwise, each disjoint seed contains at least $s+1$ mismatches, and hence the number of mismatches, m, must be greater than or equal to

$$(s+1)\left\lceil \frac{m+1}{s+1} \right\rceil,$$

which is no less than $m+1$ and is therefore a contradiction. □

Again, let l denote the seed length $\left\lfloor \frac{|Q|}{\lceil \frac{m+1}{s+1} \rceil} \right\rfloor$. Allowing at most s mismatches is effective for extending the seed length, and this should reduce the expected number of seed hits. However, this relaxation increases the number of seed search trials required to look for the positions where a seed of length l may partially match with at most s mismatches. More precisely, we generate all strings such that i ($\leq s$) nucleotide letters among l positions in the seed are replaced with the other three nucleotide letters, and use these strings as additional seeds, the total number of which is

$$3^i \cdot \binom{l}{i}.$$

Subsequently, we check whether each seed matches an off-target sequence exactly, although some probes may fail to match any sequence. As there are $\lceil \frac{m+1}{s+1} \rceil$ seeds in Q, the total number of seeds to check is

$$\left\lceil \frac{m+1}{s+1} \right\rceil \cdot \sum_{i=0}^{s} 3^i \cdot \binom{l}{i},$$

because we replace one nucleotide with the three other nucleotides. When a large number of mismatches are allowed, the number of seed search trials tends to dominate the overall performance. Conversely, as the seed length l increases, we can decrease the expected number of seed hits, i.e.,

$$\frac{|T|}{4^l} \cdot \left\lceil \frac{m+1}{s+1} \right\rceil \cdot \sum_{i=0}^{s} 3^i \cdot \binom{l}{i}$$

Example 7.6 Suppose that $|T| = 1.5 \times 10^8$, $|Q| = 19$, and $m = 4$. Table 7.4 shows the number of seed search trials, expected number of seed hits, and seed length for $s = 1, 2, 3$. Observe that allowing at most two mismatches minimizes the expected number of seed hits. Comparison to the naive and BYP methods shows a dramatic reduction in the expected number of seed hits, while allowing a slight increase in the number of seed search trials, demonstrating the effectiveness of using partially matching seeds.

Table 7.4 Comparison between exactly and partially matching seeds.

	s	number of seed search trials	expected number of seed hits	seed length
partially	1	57	2,087,402	6
matching	2	704	402,832	9
seeds	3	5,240	2,998,352	9
perfect	naive	16	39,843,750	3
match	BYP	5	11,718,750	3

Elimination of redundant seed search trials

The number of seed search trials can be further reduced.

Example 7.7 Let us consider the design of partially matching seeds when $|Q| = 19$, $m = 2$, and $s = 1$. In the previous running examples, we have set m to 4, but in this example, we set m to 2, for simplicity. Note that the seed length l equals 9 $\left(= \left\lfloor 19/ \left\lceil \frac{2+1}{1+1} \right\rceil \right\rfloor \right)$. The analysis in the previous subsection showed that the number of seed search trials is:

$$\left\lceil \frac{2+1}{1+1} \right\rceil \cdot \left\{ \binom{9}{0} + 3 \cdot \binom{9}{1} \right\} = 56.$$

However, Figure 7.15 shows that some seeds are redundant. Boxes denote substrings of Q, where long boxes represent seeds of length nine and

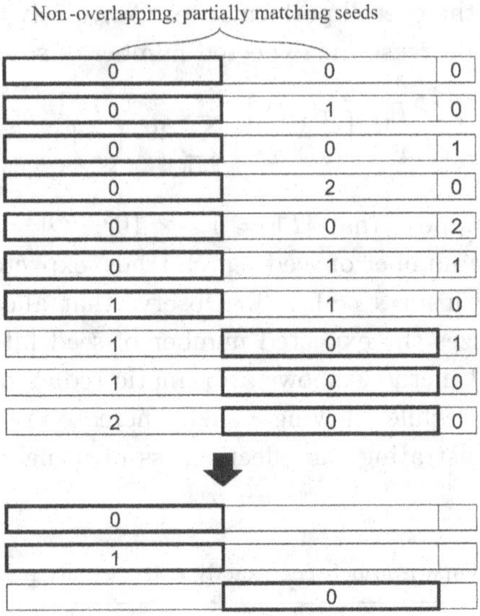

Fig. 7.15 The upper ten rows show all possible patterns, while the bottom three patterns are sufficient as seeds to search.

short boxes are of length one. Each box displays the number of mismatches. The upper ten rows enumerate all ten possible patterns of seed search trials. Each pattern can be detected by using one of the two seeds in long boxes. To reduce the number of seed search trials, we ought to select the seeds with fewer mismatches, which are enclosed in bold lines in the figure. Observe that the three patterns at the bottom are sufficient to cover all the patterns; therefore, the number of seed search trials required is

$$2 \cdot \binom{9}{0} + 3 \cdot \binom{9}{1} = 29$$

In general, Yamada and Morishita stated that the number of seed search trials has the following property [106].

Theorem 7.1 Let w denote $\left\lceil \frac{m+1}{s+1} \right\rceil$, and let $m_i (i = 1, \ldots, w)$ be the number of mismatches in the i-th seed. Then, $\sum_{i=1}^{w} m_i \leq m$, and one of the following inequalities holds:

$$m_1 \leq s, \quad \ldots, \quad m_{m+1-sw} \leq s, \quad m_{m+1-sw+1} \leq s-1, \quad \ldots, \quad m_w \leq s-1$$

Therefore, checking all the patterns of these inequalities will identify a partial seed match that meets one of the inequalities. Let l be the seed length. The number of all seed search trials is:

$$\left\{ w \sum_{i=0}^{s-1} 3^i \cdot \binom{l}{i} \right\} + (m + 1 - sw) \cdot 3^s \cdot \binom{l}{s}$$

Proof. Otherwise, all of the following inequalities hold:

$$m_1 \geq s+1, \quad \ldots, \quad m_{m+1-sw} \geq s+1, \quad m_{m+1-sw+1} \geq s, \quad \ldots, \quad m_w \geq s$$

Then,

$$\sum_{i=1}^{w} m_i \geq sw + (m + 1 - sw) = m + 1,$$

contradicting the assumption that $\sum_{i=1}^{w} m_i \leq m$.

Next, if $m_i \leq s$, the i-th seed can contain at most s mismatches, and hence the number of seed search trials using this seed is $\sum_{i=0}^{s} 3^i \cdot \binom{l}{i}$. □

7.9 Overlapping, Partially Matching Seeds

Thus far, we have assumed that seeds do not overlap. However, using nonoverlapping seeds imposes severe restrictions. For example, when $|E| = 19$, only one nonoverlapping seed with a length greater than nine can be used. What happens if we use overlapping seeds instead? Yamada and Morishita explored this approach [106].

Figure 7.16 shows the case in which the seed length l exceeds $\lfloor |Q|/2 \rfloor$. Each box shows the number of mismatches within it; we should select the seeds with fewer mismatches, which are enclosed in bold lines, to reduce the number of seed search trials. Although the upper ten seeds enumerate all of the patterns, the lower six probes are sufficient to search off-target sequences.

One may be concerned that the number of seed search trials would explode when all the mismatches are squeezed into the central box, which is illustrated in the third line from the bottom in Figure 7.16. In reality, however, the number can be kept moderately low when $|Q| = 19$. For example, if seeds of length 11 are used, the central box has three nucleotides. Selecting two positions of mismatches from the central three letters and replacing each of the two with the other three nucleotides yields the following

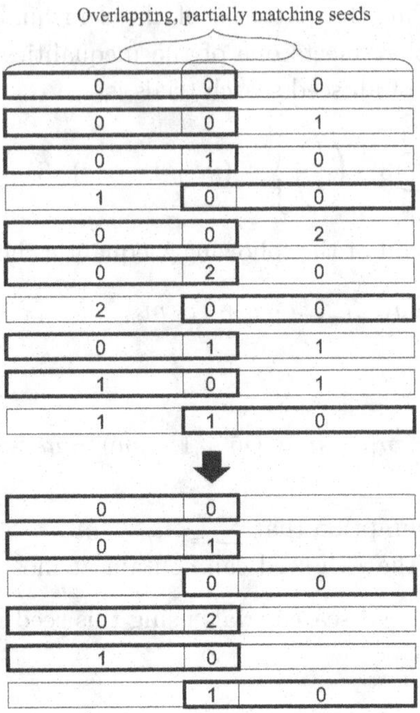

Fig. 7.16 Reducing the number of seed search trials for overlapping seeds.

number of seed search trials:

$$3^2 \binom{3}{2} = 27.$$

Here we present a simple formula to count the number of seed search trials for each pattern. Suppose that two overlapping seeds of length l share $2l - |Q|$ letters in the central interval. Let m_C denote the number of mismatches in the central $2l - |Q|$ letters. Let m_L (m_R) be the number of mismatches in the left (right) box excluding the central box. The minimum number of seed search trials for this overlapping seed is:

$$3^{m_C} \binom{2l - |Q|}{m_C} \times \min\left\{ 3^x \binom{|Q| - l}{x} \mid x = m_L, m_R \right\}$$

Summing the numbers of seed search trails for all the patterns yields the total number.

Next, we will demonstrate the power of overlapping, partially matching

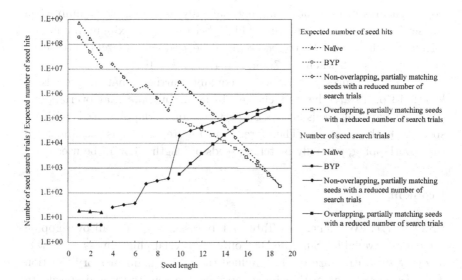

Fig. 7.17 Comparing the performance of the naive algorithm, BYP method, and two algorithms using nonoverlapping/overlapping, partially matching seeds with a reduced number of seed search trials (*Bioinformatics*, 2005, 21(8):1322).

seeds by comparing the naive algorithm, BYP method, and two algorithms that utilize nonoverlapping or overlapping partially matching seeds with a reduced number of seed search trials. We assume that $|T| = 1.5 \times 10^8$, $|Q| = 19$, and $m = 4$ because these parameters model the design of siRNA sequences for human genes. In nonoverlapping, partially matching seeds, s mismatches are allowed at most, which minimizes the expected number of seed hits, i.e., $s = 1$ for seed length $= 4 \sim 6$, $s = 2$ for seed length $= 7 \sim 9$, $s = 4$ for seed length ≥ 10.

Figure 7.17 compares the seed length (the horizontal axis), number of seed search trials, and expected number of seed hits (the vertical axis) [106]. Note that in most cases, the expected number of seed hits decreases monotonically with the seed length, while the number of seed search trials increases monotonically. Both the naive algorithm and BYP method can use seeds with a maximum length of three, so the corresponding line graphs terminate at seed length three. It is impossible to reduce the expected number of seed hits for these two methods by increasing the seed length.

In contrast, the other two algorithms can extend the seed length by using partially matching seeds, which decreases the expected number of seed hits but increases the number of seed search trials. One might wonder why both measures using nonoverlapping, partially matching seeds jump when the seed length is seven and ten. This is because the maximum number of mismatches in one seed s increases when the seed length l gets to be seven and ten; i.e., $s = 2$ when $l = 7$ and $s = 4$ when $l = 10$.

If both measures are evaluated equally, partially matching seeds of length 13 or 14 are the best choice. In addition, note that overlapping, partially matching seeds always outperform nonoverlapping seeds with respect to both measures. Therefore, from Figure 7.17, it is recommended that overlapping, partially matching seeds of length 13 or 14 be used.

Problems

Problem 7.1   We generated Table 7.1 by assuming that all overlapping k-mers are available. Suppose that only nonoverlapping k-mers that start at every k position are used to reduce the use of main memory (Section 3.5). Develop an algorithm for computing the sensitivity and specificity of finding a single hit of k-mer. Implement the algorithm, generate tables for a variety of k-mers, query lengths, and match ratios, and compare the tables with Table 7.1.

Problem 7.2   For the same project outlined in the previous problem, consider a search for multiple hits of nonoverlapping k-mer seeds, and compare the tables with Table 7.2.

Problem 7.3   Figure 7.9 displays the distribution of k-mer occurrence frequency in the human genome, which is rather different from the distribution according to the assumption that nucleotides occur at random because of repetitive sequences in the human genome. Determine this difference by generating the distribution for random nucleotide occurrences.

Problem 7.4   Figure 7.11 shows the number of k-mers of mismatch tolerance $M = 0, 1, 2, 3, 4$ in the set of human genes. Given a target nucleotide sequence of length 1.2×10^8 in which nucleotides occur at random, develop an algorithm for calculating the expected number of k-mers of mismatch tolerance $M = 0, 1, 2, \ldots$, and compare the result with the graph in Figure 7.11.

Problem 7.5   Consider the query Q of length $|Q|$ and m mismatches. Develop an algorithm that reduces the number of seed search trials when using overlapping, partially matching seeds of length l.

Chapter 8

Whole Genome Shotgun Sequencing

In 1982, Sanger et al. reported the genome sequence of the *bacteriophage lambda* using shotgun sequence assembly [86]. The number of genomic base pairs for various species can encompass a vast range. For example, *bacteriophage lambda* has 48,502 base pairs (bp), while the estimated genome size of *Escherichia coli* is about 4.6 million bp, that of budding yeast (*Saccharomyces cerevisiae*) is 12 million bp, fruit fly (*Drosophila melanogaster*) is 180 million bp, rice (*Oryza sativa*) is 430 million bp, medaka fish (*Oryzias latipes*) is 800 million bp, zebra fish (*Danio rerio*) is 1.7 billion bp, chicken (*Gallus gallus*) is 1 billion, human (*Homo sapiens*) is 3 billion, and wheat (*Triticum aestivum*) is 17 billion bp. The largest known genome is estimated to have no less than 670 billion bp (*Amoeba dubia*) [56]. If a device capable of accurately reading a long DNA sequence from one end of the chromosome to the other were to exist, sequencing a genome would be a trivial task. In reality, however, a single run of a sequencer can yield only 700-1000 consecutive nucleotides due to technical limitations. Although reading every 1000 nucleotides of a sequence would superficially appear to suffice, DNA sequence cannot be read from a specified position, as the process would require too much time and labor, which is not usually feasible for whole-genome sequencing.

How, then, a whole genome sequence be determined? One solution is the *whole genome shotgun (WGS) method*. As mentioned above, in practical terms, a DNA sequence cannot be read from a specified position; instead, reading DNA sequences from random positions on the genome is much easier and less expensive. In the whole genome shotgun method, the genome sequence is first randomly sheared into fragments, and then the fragments are sequentially read by DNA sequencers. As the number of read fragments increases, more of the fragments will originate from neigh-

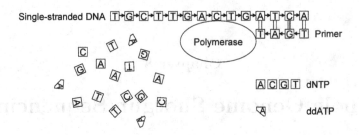

Fig. 8.1   Sequencing reaction.

boring locations on the genome. When corresponding genomic regions of the two fragments overlap, they share a common DNA sequence. When such overlapping regions are located, they can be merged into a longer sequence using a computer program called an *assembler*. When a sufficiently large number of fragments have been read, the original genome sequence can be reconstructed.

Not too long ago, few people could imagine that the day would ever arrive when mammalian genome sequences could be mapped using the shotgun sequencing approach. The invention of the automated capillary sequencer, new computer programs that can implement complex and elaborate algorithms, and other efforts to improve the performance of shotgun sequence assembly have pioneered the current state-of-the-art massive shotgun sequencing era. As the size of the target genome increases, processing a large volume of data becomes more difficult; this has fueled the search for new and more sophisticated algorithms with which to reconstruct the genome sequence. The problem of reconstruction can vary because the characteristics of a particular genome may be completely different from those of a previously sequenced species. Therefore, sequencing novel genomes demands the development of new algorithms to meet these technical, experimental, and algorithmic challenges. Whole genome shotgun sequencing has not yet become routine and remains a complex, dynamic, challenging, and exciting problem.

8.1   Sanger Method

To truly appreciate the various restrictions and limitations associated with the whole genome shotgun method, it is necessary to understand how DNA sequences are read. The basic idea for the most widely used sequencing

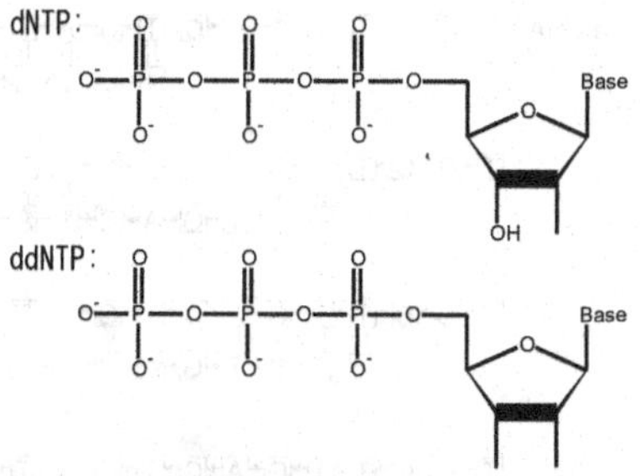

Fig. 8.2 dNTP and ddNTP.

technique was conceived by Frederick Sanger in 1975 [87]. The technique utilizes dideoxytriphosphate, and is thus called the *dideoxy method* or simply the *Sanger method*, which is described below. To read a sequence, four types of mixtures are used. One mixture contains dideoxyadenosine triphosphate (ddATP), one contains dideoxycytosine triphosphate (ddCTP), one contains dideoxyguanosine triphosphate (ddGTP), and one contains dideoxytyrosine triphosphate (ddTTP). The four mixtures also contain single-stranded template DNA, dNTPs[1], primers, and polymerases (Figure 8.1). The primers are first radioactively labeled. In the sequencing reaction, the primers hybridize to the template DNA under appropriate conditions. As occurs in the cell, the polymerases begin to replicate the template DNA, incorporating dNTPs as individual nucleotides. Because ddNTPs have a structure very similar to that of dNTPs (see Figure 8.2), the polymerases occasionally incorporate ddNTPs instead of dNTPs into the sequence, although less efficiently. Once a ddNTP is incorporated, the sequence extension stops prematurely due to the structural difference between dNTPs and ddNTPs (see Figure 8.3). The absence of an oxygen atom at the 3' carbon of the ddNTP prevents the next nucleotide from being incorporated. Most types of polymerases are capable of eliminating a base at the 3'-end of the sequence (3'-exonuclease activity) so that incorrect bases are removed and the sequence extension can proceed. The polymerases for the sequencing reaction, however, do not have 3'-exonuclease activity because it has been artificially disabled by genetic engineering. As a result, most of the long fragments in the mixtures maintain a ddNTP at the 3'-end. After eliminating the small molecules, long fragments terminated by

---

[1] dNTP represents deoxyadenosine triphosphate(dATP), dCTP, dGTP and dTTP. Similarly, ddNTP represents ddATP, ddCTP, ddGTP and ddTTP.

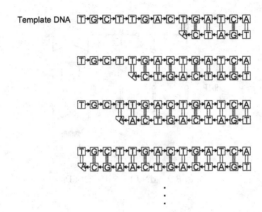

Fig. 8.3 Extension stop at 'A'.

ddATP are dominant in the mixture with ddATP.

The product is then separated using electrophoresis. When an electrical current is applied to the product, smaller fragments move faster; therefore, electrophoresis separates the fragments according to differences in lengths. Since the primer is radioactively labeled, the pattern on the X-ray film shows how long the terminated fragments are. By placing the four lanes parallel to each other (Figure 8.4), the banding pattern showing the DNA sequence can be seen.

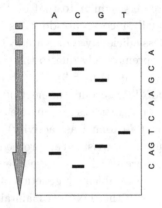

Fig. 8.4 The autoradiography shows the original DNA sequence.

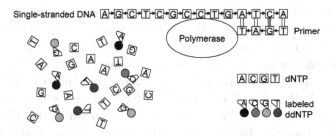

Fig. 8.5 Sequencing reaction.

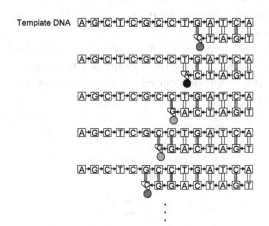

Fig. 8.6 ddNTP terminates extension.

8.1.1 Improvements to the Sequencing Method

Another breakthrough in sequencing techniques is the four-color fluorescent dye method [96]. The key difference is that this method uses four-color fluorescent dye, whereas the classical Sanger method uses radioactivity. Instead of labeling primers, each type of ddNTP is labeled with a different fluorescent dye (Figure 8.5). Electrophoresis no longer requires four lanes, and only one mixture is prepared in this sequencing method, which contains single-stranded template DNA, dNTP, ddNTP, primers, and polymerases. Another difference is that the dye method uses four types of ddNTP, i.e., ddATP, ddCTP, ddGTP and ddTTP (Figure 8.6). The polymerases extend primers to the 3'-end, occasionally incorporating colored ddNTPs. The dideoxynucleotide at the 3'-end of the fragments can be identified using laser detectors following electrophoresis (Figure 8.7). An initial problem encountered with this method was that adding different fluorescent dyes led to different electrophoretic speeds, because the dyes differ in terms of mobility during electrophoresis. For example, a sequence of 100 base pairs with ddATP at the 3'-end would be faster than a sequence of 99 base pairs with ddGTP if the fluorescent dye attached to ddATP had lower mobility than the dye attached to the ddGTP. Although this will not be discussed further, note that on a practical level, this problem has been solved. Assuming that the electrophoresis of four types of fragments is independent, the mobilities of the four types of fragments are calibrated by comparing them to reference fragments of a known size with a known type of terminal ddNTP.

Several other techniques including the use of capillaries in electrophoresis and the development of DNA polymerases that efficiently incorporate ddNTPs [99, 100], have contributed to improve the overall performance of DNA sequencers.

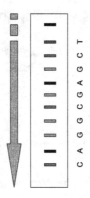

Fig. 8.7 Reading a sequence of colors is a proxy for reading a DNA sequence.

8.2 Cloning Genomic DNA Fragments

Although the improvements described in the previous section have eliminated various restrictions related to DNA sequencing, the current best sequencing still requires following conditions:

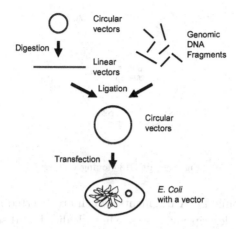

Fig. 8.8 Genomic DNA fragments are inserted into vectors, which then transfect the host bacteria.

(1) The amount of template DNA must be appropriate (e.g., 0.1-3.0μg). Because a sequencer "sees" nucleotides via the fluorescent dye, determining the type of nucleotide cannot be performed with a single copy of the DNA fragment. Multiple copies of the same fragment must be prepared so that the laser detector can robustly identify the nucleotides.
(2) The template DNA must be single-stranded. In typical sequencing protocols, double-stranded DNA is denatured under appropriate thermal conditions.
(3) Primers are required to initiate the sequencing reaction. Recall that DNA sequences cannot be read from arbitrary positions.

Whole genome shotgun sequencing requires sheared short fragments, each of which is only a single copy of the DNA molecule prior to amplification. To meet these conditions, sheared DNA fragments are amplified by *cloning*, an extremely important genetic engineering technique. Cloning vectors can accommodate foreign DNA fragments and are usually hosted by the bacterium *E. coli*. As the host *E. coli* proliferates, the cloning vectors are replicated along with its nuclear genome. Typical cloning vectors, which have a circular form, are first prepared and digested by restriction enzymes, and then the digested vectors are mixed with sheared DNA fragments. After the ligation, the vectors containing the genomic DNA fragments are transfected to host *E. coli* (Figure 8.8). Subsequently, they are planted on an appropriate culture medium. The colonies that form provide multiple

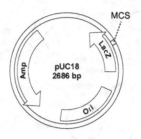

Fig. 8.9 pUC18 cloning vector.

copies of the genomic DNA fragments, which are inserted into the vectors.

However, this description is somewhat idealized, and several problems can be encountered as follows.

- Not all of the vectors have taken genomic DNA fragments in. Although it depends on the protocol, only a small fraction of the vectors may contain genomic DNA fragments.
- Also, depending on the protocol, most of the *E. coli* may remain untransfected.
- Sometimes more than one fragment is inserted into the vectors; i.e., two independent genomic DNA fragments may both be inserted into the vector at the same time. Because these vectors have DNA sequences that originate from different parts of the genome, they are called *chimeric clones*.

Several tricks are used to overcome these problems.

The typical cloning vector pUC18 is illustrated in Figure 8.9. "Ori" represents a replication origin, from which replication begins in the host *E. coli*, and "Amp" represents an ampicillin-resistant gene. When pUC18 is transfected into *E. coli*, most of the bacteria remain untransfected. To remove untransfected *E. coli*, they are cultured in a medium with the antibiotic ampicillin. Because transfected *E. coli* contain the vector that carries the ampicillin-resistant gene, only *E. coli* with the vector can survive; untransfected *E. coli* cannot form a colony.

Another trick involves a multicloning site (MCS) located in the middle of the LacZ gene on pUC18. MCS is a sequence that contains several recognition sequences of restriction enzymes. For example, the MCS of pUC18 contains *EcoRI*, *SacI*, *KpnI*, *BamHI*, *XbaI*, *SalI*, *PstI*, *SphI*, and *HindIII* sites. The vector is cut at the MCS, and genomic DNA fragments are inserted at the cut site. LacZ on pUC18 has an appropriate promoter sequence so that the LacZ gene is expressed in *E. coli*. LacZ codes for β-galactosidase, which hydrolyzes 5-bromo-4-chloro-3-indoyl-B-D-galactoside (X-gal). X-gal, which is colorless but becomes blue in the presence of β-galactosidase, can be added to the medium to detect the presence of LacZ. When a genomic DNA fragment is inserted in the middle of the LacZ gene, the gene is disrupted. A blue colony indicates that the *E. coli* contains a vector without fragments inserted, whereas a white colony signifies that the *E. coli* contains a vector with a fragment. Therefore, nonrecombinant clones can be distinguished from recombinant clones by the colors of the colonies. Colony selection is now fully automated; a CCD camera scans the plate, and then the white, well-isolated colonies are selected by a dedicated instrument.

When a sequencing template is needed, vectors can be isolated from the host *E. coli*. Isolated vectors are then usually amplified to the appropriate amount by rolling circle amplification [9, 24, 29, 57, 82].

8.3 Basecalling

DNA sequencers output four-color waveforms obtained through electrophoresis. Because raw waveforms are not suitable for reading DNA sequences, several calibrations must be performed, one of which is the mobility calibration described in the previous section; another is cross-talk calibration. Detectors of the four colors can cross-talk because the fluorescence intensity spectra of the four dyes overlap. Other calibration methods include peak height calibration and peak spacing calibration. After the calibrations are completed, the analysis can proceed with the *electropherogram*, shown in Figure 8.10.

The next task is to convert an electropherogram into a series of nu-

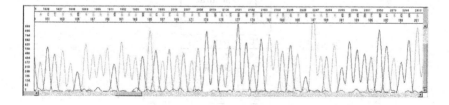

Fig. 8.10 Electropherogram of a read from the medaka WGS project.

cleotides, i.e., A, C, G, and T. The conversion process is called *basecalling*, which is often accomplished using the computer program *phred* [32]. Phred detects a series of peaks in input electropherograms, and outputs nucleotide sequences and their *quality values* (QVs), which use a logarithmic scale and satisfy the following equation:

$$QV = -10 \log_{10} P,$$

where P is the probability of an error at a base. A QV is assigned to each base, indicating the probability of an error at that base. For example, if the QV of a particular base is 10, the probability of error is 10%. A QV of 20, which corresponds to 99% accuracy, is often used as the cutoff threshold for various analyses, and a QV of over 50 indicates "almost correct" in practice. The relationship between the QV and the accuracy of bases is shown in Table 8.1.

The QVs that are determined represent experimental statistics. A DNA sequence with a known sequence is prepared and sequenced several times. Five parameters are calculated for each called base:

- the ratio of the longest spacing between peaks to the shortest spacing between peaks in the 3-base window;
- the ratio of the longest spacing between peaks to the shortest spacing between peaks in the 5-base window;
- the ratio of the height of a called peak to the highest uncalled peak in the 3-base window;
- the ratio of the height of a called peak to the highest uncalled peak in the 5-base window;
- the number of bases to the nearest "N" in the called sequence.

The called bases are plotted in five-dimensional space. A heuristic algo-

Table 8.1 The relationship between the QV and basecalling accuracy.

QV	accuracy
4	60%
7	80%
10	90%
14	96.0%
20	99.0%
24	99.6%
30	99.9%
40	99.99%
50	99.999%

rithm is used to partition the space according to every QV based on the basecalling accuracy of the points. For example, a subspace of the space is assigned QV 20 because the basecalling accuracy in the subspace was 99%. The five parameters are also calculated for each base of a new sequence. Assuming that the statistical distribution of the points in the five-dimensional space is the same as the reference experiment above, the QV of each base is determined by the subspace containing the five parameters.

Note that a QV cannot always be relied upon when determining whether a particular base mismatches other bases. Moreover, two bases in the same read are strongly correlated in terms of accuracy, and the QVs of two bases on two separate reads may be correlated in some cases. If an extremely large number of bases with QV 20 is present, about 99% may be correct. However, when two reads are aligned and a particular mismatched base of the reads has a QV over 20, the probability of sequencing error at that base is not always lower than $1\% \times 1\% = 0.01\%$.

8.4 Overview of Shotgun Sequencing

We have come a long way from sheared genomic DNA fragments to nucleotide sequences with QVs. Based on the techniques described above, following is a brief overview of shotgun sequencing.

In the whole genome shotgun method, multiple copies of genomic DNA are prepared (Figure 8.11a). The genome DNA is sheared randomly by hydrodynamic shearing forces, i.e., it is sheared by fast water flow (Figure 8.11b), and subsequently, the fragments are fractionated in size (e.g., 4,000 bp) (Figure 8.11c). Electrophoresis, sucrose density gradient centrifugation, or other techniques are used for the size fractionation. The fractionated DNA fragments are inserted into cloning vectors for amplification. Several types of cloning vectors are available, and individual cloning vectors can accept fragments of different sizes. Plasmids are typical cloning vectors, and they can host fragments of up to 300,000bp, although the size depends on the type of vector used. The maximum size of acceptable fragments is limited by biological mechanisms. High-copy plasmids, such as pUC18, accepts relatively short fragments (e.g. <20,000bp), whereas a single-copy Bacterial Artificial Chromosome (BAC) accepts fragments of up to 300,000bp. The fosmid vector, which is an improvement on the cosmid vector, is another vector that accepts fragments of 25,000-45,000bp. Although the other types of vectors usually have no lower limit to the size of the acceptable fragments, the fosmid vector cannot accept shorter fragments because of its unique biological packaging process. When typical high-copy plasmids are used, fragments of less than 2,000 bp are preferable in terms of cloning efficiency; however, cloning fragments of less than 2,000 bp are ineffective for shotgun sequencing. In a single run of a DNA sequencer, a DNA sequence of up to 1,000 bp, termed a *read*, can be obtained from a primer site. Typically, the whole vector sequence is known, and therefore, both ends of an inserted fragment can be read starting from the primer sites at the two sides of the inserted fragment. If the size of a fragment is less than 2,000 bp, two sequences read from the primer sites overlap in the middle of the inserted fragment, which leads to a reduced effective sequence coverage of the genome. Without size selection, shorter fragments are selectively cloned because they are more efficiently cloned than longer fragments; therefore, size selection is critical to avoid such inefficiency.

The black lines in Figure 8.11d represent reads, and the gray lines represent the parts of the fragments that are not included in the reads. Some of the fragments do not contain gray lines because two reads in these fragments have overlapped, and no parts of the fragments remain unread. The typical occurrence of such fragments is < 3%. Some of the fragments are read at only one end because a good quality read cannot be obtained at the other end; the typical occurrence of such fragments is < 5%. Some of the fragments are completely unreadable, so that no reads can be obtained from them, which is not shown in the figure. The typical occurrence of such fragments ranges from 3% to 20%, depending on the protocol and the charcteristics of the target genome.

To reconstruct the original genome sequence, short reads that overlap must be assembled to yield longer sequences. Because the sequence overlaps rely heavily on the stochastic process of fragment sampling, the obtained read sequences may not be sufficient to reconstruct the original genome sequence. There may be many *stochastic gaps*, i.e., some of the genomic regions may not be covered by any reads. Other types of gaps will be described later. Overlapping reads are merged to form several islands, or contiguous sequences termed *contigs*, which are denoted by the first six letters in the contiguous sequence (Figure 8.11e). This figure shows nine contigs, each of which consists of multiple reads.

A higher-order structure that orders and orients the contigs can then be created. A pair of reads from one fragment is called a *mate pair* (Figure 8.12), which provides information about the relative order and orientation of the two reads. The two reads originate from opposite strands on the

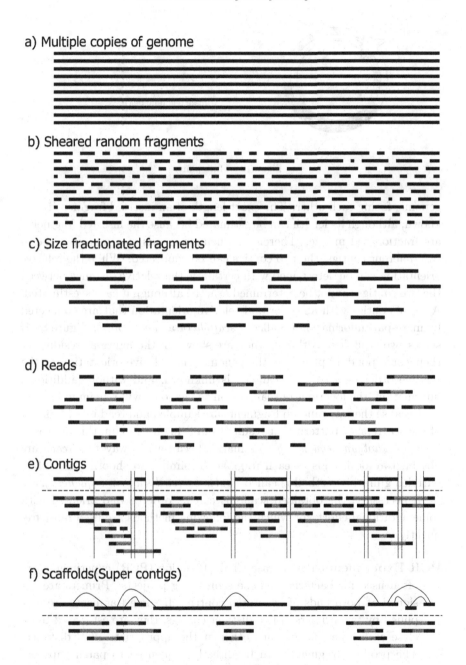

Fig. 8.11 Shotgun sequencing.

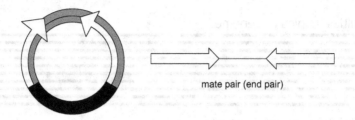

Fig. 8.12 Mate pair.

genome, facing each other. The exact length of an inserted fragment is not known, although its length can be estimated because the inserted fragments are fractionated in size. Therefore, when two reads of a mate pair lie in two different contigs, the two contigs can be connected. While the relative orientation can be determined with certainty, the relative distance between the two contigs cannot be determined exactly, although it can be estimated. A set of contigs with an order and relative orientation that are connected by mate-pair information is called a *scaffold* or a *super contig*. Figure 8.11f shows two scaffolds. Although they are shown in the figure as residing in their corresponding places on the genome, the relative orientation or the relative order of a pair of scaffolds is unknown, and without additional information, the locations of scaffolds on the genome will remain unknown.

Because the way random fragments are sampled and read is reminiscent of the scattershot pattern that results from firing a shotgun, this approach is called *shotgun sequencing*. Similarly, because the way two *reads* are obtained at both ends of each fragment is similar to shooting a double-barreled gun, this method is called *double-barreled shotgun sequencing*.

If necessary, the obtained scaffolds can be ordered and oriented with some additional effort, for which the following methods are used most frequently.

PCR Experiments Polymerase Chain Reaction (PCR) experiments are often used for bacterial genome sequencing projects. Primers are designed at both ends of generated contigs. If some fragments are amplified by two primers of two distinct contigs, they are estimated to be close each other, facing each other at the appropriate PCR distance. The product fragments can be directly sequenced to patch gaps between contigs. However, when the number of contigs is too large, this approach is not feasible because the number of pairs of contigs is too large even if multiplex PCR is adopted to optimize the number of PCR experiments.

Genetic Map If a target species has historically been used in genetic studies, a genetic map is likely to exist. Scaffolds can be related to the genetic map by searching for the markers on the scaffolds, although some types of markers (e.g., phenotype markers) are not suitable for anchoring scaffolds. For example, in a sample of 10,000 sequence tagged site (STS) markers whose relative genetic distances are known, at most 10,000 scaffolds can be ordered in the linkage groups. Note that at least two distant markers are required to orient a scaffold; otherwise, several scaffolds may form a cluster, or it can be ordered but not oriented.

BAC Contigs If BAC libraries are available, they can be arranged using BAC fingerprinting. The restriction patterns of the BACs are determined, after which overlapping BACs are connected by shared bands. A group of the connected BACs are called a BAC contig. The end sequences of BACs are usually used to anchor scaffolds on BAC contigs. Other techniques may be also used for the construction of BAC contigs.

Radiation Hybrid Map The term "radiation hybrid"(RH) refers to a hybrid between rodent cells and the cells of target species. The cells of target species are terminally irradiated, after which they are fused with the rodent cells. As some of the fused cells have extra DNA fragments that originate from the target species, these cells can be used for mapping. For example, both two pairs of primers that amplify a unique region of the target species yield a band for the radiation hybrid, it is suggested that the two loci are relatively close to each other because distant loci are more unlikely to slip into the RH cell at once. However, irradiation sometimes causes unknown behavior, making radiation hybrid mapping less reliable than other procedures.

Optical Map Optical mapping is an innovative mapping technique, in which a single molecule of a stretched chromosome is first fixed to solid surface. The chromosome is subsequently cleaved by restriction enzymes, and a photograph of the restricted chromosome is taken with a CCD camera. The picture is analyzed by computer software, and a comprehensive restriction map of the chromosome can then be obtained. The accuracy of the measured distances between cleavages is <5% [89]. The created map is called an *optical map*. Comparing an *in silico* restriction map of scaffolds to an optical map enables us to order and orient scaffolds.

Fluorescence in situ Hybridization (FISH) FISH is one of the techniques used to determine where a particular probe can hybridize in the genome. Fluorescently-labeled probes are hybridized to the genome *in situ*, and the colors that are produced indicate how the probes are ordered. The precise distance between probes is unknown, because chromosomes are not stretched straight as in the optical map. The physical distances between probes do not have a strictly linear relationship with the number of bases between probes. Large differences, such as 800 kb and 80 kb, are easily distinguishable, but it may not be possible to distinguish 800 kb from 700 kb.

Synteny Information Synteny information is fundamentally different from the other experiments described above. If two species, X and Y, are evolutionarily close to each other, their genome sequences are expected to have a similar structure. When the genome sequence of species X has already been elucidated, it can be used as a template for species Y. Shotgun reads of Y are first assembled to scaffolds. A simple homology search then allows us to map the scaffolds to the genome sequence of X as long as they have similar nucleotide sequences. After mapping, we can order and orient the scaffolds of Y, assuming both X and Y have the same genome structure. Of course, this strategy carries the risk of misordering scaffolds of species Y due to the genomic rearrangement between X and Y. Therefore, syntenic information should be used to support weak links produced by other experiments. When the genome sequences of X and Y are nearly identical, the shotgun reads can be directly mapped onto the genome of X, constructing contigs [78]. This method is also useful when the number of the shotgun reads is so small that they do not hold sufficient information to reconstruct the whole genome.

8.5 Lander-Waterman Statistics

Under ideal conditions, it is possible to calculate the expected number and the average length of contigs [53]. Given a genome of size G, suppose that individual fragments are sampled at random positions uniformly over the genome, and all consist of the same number of base pairs, e.g., L base pairs. To simplify the problem, assume that only one end of each inserted fragment is read, unlike the usual double-barreled shotgun sequencing. In other words, the positions at which the fragments are read are entirely independent of each other. Two fragments that share a common substring of length no less than $L\theta$ ($0 \leq \theta \leq 1$) bp are then joined at the end of one fragment and at the head of the other. If N is the number of fragments, the following property emerges:

Theorem 8.1 *The expected number of contigs is $Ne^{-(1-\theta)\frac{LN}{G}}$.*

Proof. A contig stops at the last overlapping fragment with no following fragments. Thus, the expected number of contigs coincides with the expected number of such "stopper" fragments. In a stopper fragment, no fragments appear at any of the first $L(1-\theta)$ base pairs. The probability that some fragments appear at an arbitrary position is N/G, according to the ideal assumption that each fragment occurs at a random position. Therefore, the probability of having a stopper fragment is

$$\left(1 - \frac{N}{G}\right)^{L(1-\theta)} = \left\{1 + \left(-\frac{N}{G}\right)\right\}^{\frac{1}{(-\frac{N}{G})} \cdot -\frac{LN}{G}(1-\theta)} \approx e^{-\frac{LN}{G}(1-\theta)}$$

because $-\frac{N}{G} \approx 0$. Hence, the expected number of contigs is $Ne^{-(1-\theta)\frac{LN}{G}}$ □

$\frac{LN}{G}$ expresses the ratio of the total length of fragments to the size of the genome, which is the average number of fragments that covers an arbitrary position in the genome. The figure is therefore called the *sequence coverage* or *sequence cover ratio*. In practice, L ranges from 500 to 1,000, and the typical value for θ is 0.1, which implies $e^{-(1-\theta)} \approx 0.4$. r denotes the sequence coverage $\frac{LN}{G}$. The expected number of contigs is $\frac{G}{L}r0.4^r$, which almost exponentially decreases with the increase in the sequence coverage if G and L are treated as constants. In other words, the average contig size increases almost exponentially with sequence coverage, as shown in Figure 8.13. In reality, however, the average contig size is much lower than the above theoretical estimate due to a number of repetitive sequences and sequences that are difficult to read.

In 1982, Sanger et al. succeeded in elucidating 48,502 base pairs of the circular *bacteriophage lambda* genome by dividing the genome into short fragments of 250 to 300 bp and joining these fragments [86]. The report has a 19-page section that displays the sequences of all base pairs and gene coding regions. The genome length that can currently be sequenced has expanded markedly in the last 20 years; however, there has been little progress in the number of readable base pairs obtained during one sequencing reaction. Thus, the concept of shotgun sequencing remains valuable even today. The processing of vertebrate genomes requires tens of millions of fragments to be read; e.g., 33.5 million fragments were processed to sequence the

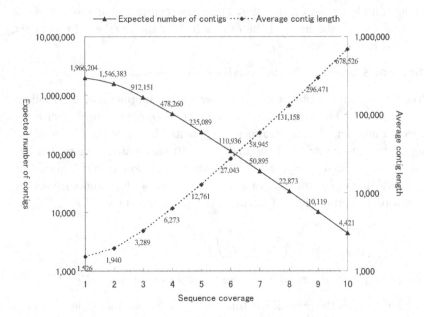

Fig. 8.13 The expected number of contigs and the average contig size for various sequence coverages when $G = 3 \times 10^9, xL = 600$, and $\theta = 0.1$.

mouse genome [21]. Manual efforts to handle such a large amount of data would be impossible, and to this end, automatic processing using computer software is indispensable.

8.6 Double-Stranded Assembly

One fragment in a shotgun read comes from either strand of the genome, but it is impossible to determine which one is the actual strand. Both cases must therefore be considered, but examining two cases simultaneously requires a complex implementation of assembly algorithms. For example, even assuming that there are no sequencing errors in the shotgun reads and the length of the reads is a fixed number, the relationships between two shotgun reads may fit no less than six cases (Figure 8.14). Because the source code should not become cluttered with deeply nested "if"s throughout, trick can be used to circumvent this problem.

Consider Figure 8.15. Assume we have the shotgun reads illustrated in the upper left of the figure. The usual assembly process may yield the

Whole Genome Shotgun Sequencing

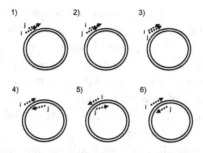

Fig. 8.14 Relationships between two shotgun reads of a fixed length.

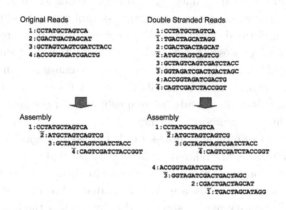

Fig. 8.15 Double-stranded assembly yields two contigs, which are reverse complementary to each other.

result shown in the lower left. Each read in the assembly is either reverse complemented or not, and the bars on both 2 and 4 indicate that the reads are reverse complemented. We define *double-stranded reads* as a set of reads that contains both the original shotgun reads and their reverse complements. For the set R, which consists of the original shotgun reads ($R = \{r_0, r_1, \ldots, r_{n-1}\}$), we define the set of double-stranded reads, $R' = \{r'_0, r'_1, \ldots, r'_{2n-1}\}$ as $r'_{2i} = r_i, r'_{2i+1} = $ the reverse complement of r_i. The double-stranded reads of the original reads in the figure are shown in the upper right. As a result of doubling the reads, the assembly result contains redundant islands of reads. The reconstruction is performed simultaneously for two strands of the original genome sequence.

The introduction of double-stranded reads may appear to have complicated the problem, but it has a distinct advantage, i.e., there is no longer a need for considering the reverse complements of the reads. Reads can simply be extended by finding shared substrings between the reads as if they were the usual text strings, with few errors. Thus, the algorithm is greatly simplified using the concept of double-stranded reads.

Nevertheless, issues of concern remain, such as additional memory consumption, the loss of complementary information, and redundant outputs. These issues are addressed below.

First, does doubling the shotgun reads double memory consumption? While it is true that naive implementation doubles memory consumption, most of the algorithms run faster with double-stranded reads because the transformation occurs only once. The shotgun reads are usually accessed many times in the whole genome shotgun assembly, therefore, if the transformation is performed every time the reverse complement reads are accessed, the total processing time will be longer than implementing the double-stranded reads. Even when the memory consumption is the bottleneck of the algorithm, it is possible to use a proxy class to access to the reverse complemented reads, accomplishing both compact description of the algorithm and memory efficiency.

Second, will assembling double-stranded reads disregard that the two complementary reads are related? The answer is no. Given the definition above, the two complementary reads maintain their relation because the index numbers of the reads are deeply related. We can determine the reverse complementary read of some read r'_i by flipping the least important bit of the index number. For a double-stranded read r'_i, r'_j is the reverse complementary read, where j = i ^ 1 in popular programming languages such as C/C++/Java/Perl. Also, we can determine that the two reads are complementary by the condition j ^ i == 1. If we wish to determine the index number of the original read, we can simply ignore the least important bit of the index number by shifting it by one bit. For a double-stranded read r'_i, r_j is the corresponding original shotgun read, where j = i >> 1. It is important to note that the read numbering scheme here is extremely efficient in run-time processing speed because the instructions above are so primitive to modern CPUs that it may require only a single clock cycle or less.

Third, is the output redundant because two reverse complementary sequences are output for each contiguous sequence? Redundancy can easily be removed. As described above, the relation between the original shotgun reads and their double-stranded reads are retained; therefore, every pair that contains reverse complementary sequences can be determined, as long as the symmetry is not destroyed during assembly.

The concept of double-stranded reads leads to both faster assembly and a simpler algorithm.

8.7 Overlap-Layout-Consensus

The practical approach that most state-of-the-art assembly programs adopt, either explicitly or implicitly, is the *overlap-layout-consensus* strategy [46, 66, 73], which consists of three major steps: detection of the overlapping fragments (overlap), layout of fragments (layout), and calculation of the consensus sequence of individual nucleotide letters from the ordered fragments in the previous step (consensus), as illustrated in Figure 8.16. The three major steps are outlined below.

8.7.1 *Overlap*

This step enumerates all fragment pairs that are likely to overlap with each other at the end of one fragment and at the head of the other. An *overlap*

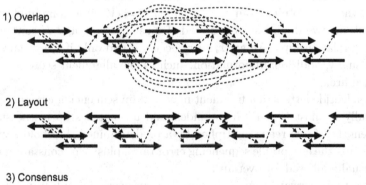

Fig. 8.16 Overlap-layout-consensus strategy.

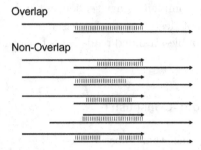

Fig. 8.17 Overlap versus non-overlap. Vertical lines indicate where two reads are well-aligned.

is defined in Figure 8.17. The vertical lines show where the two sequences are well-aligned. If two sequences are parallel to each other but there are no vertical lines, the sequences are not well-aligned. If a read r_1 *overlaps* with another read r_2, they are well-aligned if and only if the tail of r_1 is aligned to the head of r_2. Each read must not be entirely included in the other read. In the other cases of "Non-Overlap" in Figure 8.17, it is inferred that the two reads come from different positions on the genome with the exception that the last but one case indicates inclusion.

In the overlap phase, an *overlap graph* $G = (V, E)$ is computed, where V contains two double-stranded reads for each read r, and E contains all overlap relationships. An arc $r_i \to r_j$ occurs if and only if r_i overlaps with r_j. Usually, additional information, such as the alignment score, is added to each arc.

Nucleotide letters in a fragment may contain sequencing errors that are typically at most 5%. Both the indels and mismatches up to the sum of the lengths of two reads multiplied by a predetermined threshold ϵ, which represents the acceptable sequencing error ratio, plus some constant margin are usually allowed for overlaps.

A straightforward way of constructing an overlap graph from shotgun reads is to compute the overlap alignments for every pair of reads. However, if a large volume of fragments must be processed, e.g., more than 10 million for sequencing vertebrates, the number of possible fragment pairs amounts to more than 1 trillion, making the overlap detection step dominant throughout the overall computation time of the genome assembly. Thus, it is crucial to accelerate this overlap detection.

One common approach used to solve this issue is to build a lookup table that stores the locations in fragments of short strings ranging from 16 to 32bp in length. The lookup table should be in the computer memory, if possible, while it is usually processed on hard drives, or possibly on distributed clusters. One fragment can then be scanned to find other fragments that share common substrings by querying substrings in the lookup table.

This filtering step dramatically reduces the number of possible overlapping fragment pairs; however, some fragment pairs are false-positive in the sense that they do not actually overlap on the genome. Therefore, whether a pair truly overlaps needs to be precisely confirmed. To avoid checking a large number of false-positive pairs, k-mers that are highly represented in shotgun reads are usually masked. The mask is especially important for assembling shotgun reads of species equivalent to or more complex than vertebrates. For example, the shared 16-mer between two reads indicates a true overlap at over 90% probability in an average bacterium sequencing project, while the probability is much less than 0.1% in vertebrate sequencing projects because vertebrate genomes have more copies of repetitive elements. Without masks, even 100 CPU months would not suffice to complete the overlap graph for 10 million shotgun reads of vertebrate genomes unless other filtering procedures are performed.

8.7.2 *Layout*

This step attempts to derive a subset of the overlap graph that orders shotgun reads without conflict. Assuming the absence of highly similar repetitive sequences, all reads that overlap originate from some continuous region of the genome. However, repetitive sequences in the genome actually cause conflicting overlaps in the overlap graph (Figure 8.18). For example, it is possible that read A can be extended in the 3' direction by two other reads, B and C, while B and C apparently come from different positions on the genome (Figure 8.19). The 3'-end of A, the 5'-end of B, and the 5'-end of C can share the same DNA string when A is in the repetitive region of the genome. Such conflicting overlaps must be resolved to create a layout of reads, i.e., the shotgun reads must be linearly ordered.

In some cases, conflicting overlaps cannot be resolved. A typical example is illustrated in Figure 8.20. A, B, C, and D are unique regions, while R is a repetitive sequence that appears twice at different positions in the genome. When the length of R is much greater than the length of the shotgun reads, the linkage between A, B, C, and D is unknown without other information in addition to the DNA sequences. Because both

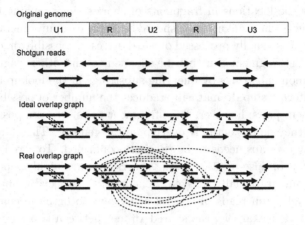

Fig. 8.18 Genome shotgun reads, the corresponding ideal overlap graph, and the real overlap graph.

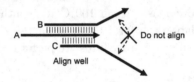

Fig. 8.19 Read A overlaps with reads B and C, while B and C conflict.

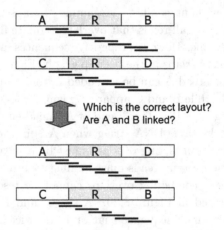

Fig. 8.20 The same set of shotgun reads can be obtained from two different genome sequences.

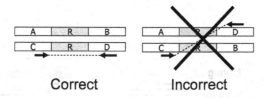

Fig. 8.21 A mate pair can provide a link over a repetitive sequence.

possible layouts of the original genome sequence shown in the figure can yield the same set of shotgun reads, the two cases cannot be distinguished. To avoid misjoining reads, a contig may be terminated at a read having conflicting overlaps, although this conservative approach is likely to yield short contigs.

Additional information is needed to further extend contigs confidently, and the major source of such additional information is mate-pair information. Because a mate pair is a pair of shotgun reads from both ends of a clone, the distance between the pair of reads on the assembly must be equal to the length of the inserted DNA fragment. Although the length of the inserted DNA fragment cannot be measured exactly, its length can be estimated because the DNA fragments were fractionated prior to cloning. If two reads of a mate pair can be placed in C and D as shown in Figure 8.21, it is probable that the left layout in the figure may be correct as long as an appropriate distance constraint is met. Two reads of a mate pair must face each other, and the distance between the two reads should be around the fractionated size. Thus, provided that the estimated length of the insert DNA is far less than the length of R+B+A, the left layout in the figure has a high probability of being correct. Note that if the estimated length of the insert DNA is much longer, the layout CRB...ARD is a possible choice, although the ambiguity remains. The key point is that *both* reads of a mate pair must originate from a unique region on the genome. Even if one read of a mate pair originates from a unique region, the other read that originates from the repetitive sequence cannot usually resolve the ambiguity. A case in which both reads of a mate pair originate from a repetitive region of the genome cannot be resolved. It should be noted again that both reads of a mate pair must originate from a unique region on the genome in order to resolve the ambiguity. This process is called *untangling* because it resolves tangles caused by repetitive sequences in an overlap graph.

Another use of mate pairs is scaffolding, in which the set of contigs is

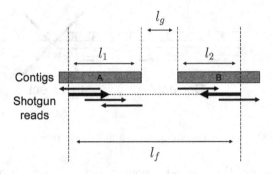

Fig. 8.22 A mate pair can provide an estimate of the gap size between two contigs.

ordered and oriented based on mate-pair information. Assume two shotgun reads from a mate pair assembled in contig A and B, respectively, as shown in Figure 8.22. Because the two reads of the mate pair must face each other, the relative orientation of contig A and contig B can be determined immediately. The relative position of the two contigs can also be estimated. First, l_f, the length of the inserted DNA fragment, can be estimated by the fractionation size. Both l_1 and l_2 can be measured precisely because the positions of the two reads on contig A/B are already known and because the sequences of the contigs are available. Provided l_f, l_1, and l_2, a gap size l_g can be calculated directly by solving the elementary equation $l_f = l_1 + l_g + l_2$. l_f, the length of insert DNA fragments, has a given distribution; therefore, l_g may also have the same distribution. The variance of l_g is not very small under typical experimental protocols; thus, l_g can often be a negative value, which violates physical constraints. The typical heuristic for this problem is to set l_g to some small fixed constant c_{min} when l_g is less than c_{min}. To accurately estimate the gap sizes, multiple mate pairs are required to decrease their variances. A set of such related contigs is called a *scaffold*, and constructing scaffolds is called *scaffolding*.

8.7.3 Consensus

After completion of the layout phase, a series of ordered shotgun reads is obtained. The final task is to create consensus sequences from the shotgun reads using multiple alignment. Because shotgun reads contain sequencing errors, some sequences disagree at the nucleotide level. However, most sequencing errors occur independently, so that the effect of sequencing er-

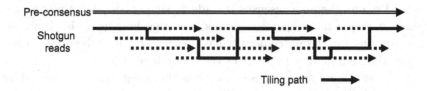

Fig. 8.23 "Pre-consensus" is computed as a tiling path (tiling bases) of shotgun reads.

rors can be eliminated by multiple alignment (Figure 8.16, Consensus). The principle invoked here is the majority vote. Although a singleton nucleotide may be an artifact of a sequencing error, a nucleotide supported by multiple instances of reads is unlikely to have resulted from sequencing errors. After multiple alignment of ordered reads, a representative sequence, called a *consensus sequence*, or a *consensus*, is determined.

Traditionally, multiple alignment algorithms have been used to align evolutionarily related protein/DNA sequences. In traditional algorithms, multiple alignment is formulated to solve the optimization problem regarding some objective function, such as the "sum of pairs." In a typical multiple alignment of proteins, amino acid sequences are similar at a relatively low homology ratio, which requires various sensitive and time-consuming computations. In contrast, shotgun reads originate from the same genome and have a much higher similarity than that encountered in traditional multiple alignment problems, suggesting that the multiple alignment of shotgun reads is easier to solve. While it is true that the problem is less concerning in the sense that the similarity between shotgun reads is high (typically $> 90\%$), the number of shotgun reads involved in the multiple alignment phase is overwhelmingly larger than in traditional multiple alignment. Therefore, traditional multiple alignment software such as ClustalW [101] cannot be applied directly to the multiple alignment of shotgun reads, and guided tree heuristics or iterative improvement heuristics take too long for aligning tens of millions of shotgun reads.

A typical approach for solving the processing time problem is the use of "Master-Slave alignment" heuristics. First, a tiling path of ordered shotgun reads is computed for each contig (Figure 8.23). The algorithms used to compute the tiling path vary between implementations, although the principle is that the bases of higher quality values are used for the tiling path. Second, a "pre-consensus," which is a tentative consensus sequence for the contig, is created as a patchwork of the shotgun reads according to the tiling path. Because the pre-consensus is made by simply concatenating parts of reads, it may retain sequencing errors from the shotgun reads. However, the pre-consensus sequence usually has over 99% accuracy because it comprises the high quality portions of the shotgun reads. In the presence of the almost correct pre-consensus, multiple alignment can be computed by a series of pairwise alignments. Every shotgun read is aligned in a pairwise manner to the pre-consensus sequence. Most of the mismatched bases are singletons that originate from sequencing errors, while multiple "votes" against the pre-consensus indicate the errors in the pre-consensus.

As may be deduced from the basecalling process described above, the pre-consensus rarely has consecutive insertion/deletion errors of more than one base. Since peak spacing in electropherograms is relatively even, three peaks are rarely confused with one peak in high quality parts of reads, although it is possible for two peaks to be confused with a single peak. Hence, an iteration rarely leads to the wrong consensus sequences, with the caveat that "rarely" does not mean "never."

This multiple alignment algorithm, which is called the *Master-Slave alignment*, first creates one tentative consensus to which the shotgun reads are then aligned.

8.8 Practical Whole Genome Shotgun Assembly

We have thus far explored the somewhat theoretical world of whole genome shotgun assembly. A combination of previously described techniques can successfully assemble bacterium genomes with some additional manual effort, but the large-scale whole genome shotgun assembly involves several issues that are not encountered in smaller projects. In the real world of whole genome shotgun assembly, many factors prevent naive processing of the shotgun reads. For example, a set of whole genome shotgun reads contains a great deal of noise attributable to experimental limitations, biological factors, and human error. Some problems would be relatively easy to solve if a trained person, rather than a machine, were able to tackle the issues. Unfortunately, no human can process tens of millions of shotgun reads within several months without making errors. In this situation, the problem is whether the behavior of a human being can be implemented and mimicked without dramatically increasing the processing time. However, even specialists would be confronted with major difficulties due to the limited amount of information that can be obtained from a set of shotgun reads. This problem cannot be solved with a high degree of confidence, which may have to be explicitly indicated in the assembly output, or additional experiments must be designed to resolve the ambiguity.

Whole genome shotgun assembly of vertebrate/mammalian genomes poses exciting but considerable challenges. These problems and their solutions are discussed below.

8.8.1 Vector Masking

Since primer sites are designed from known vector sequences, raw shotgun reads inherently contain part of the vector sequence. Figure 8.24a/b shows typical raw shotgun reads. Basically, every shotgun read has part of the vector sequence around the cloning site at its 5'-end. When the size of insert fragments is short, the other end of the vector sequence across the cloning site appears in the tail of the raw read as shown in Figure 8.24c. Because shorter fragments in the sequencing reaction include several types of contamination, the bases immediately following the primer site cannot be used. For this reason, when the primer site is designed at a sufficiently close position to the cloning site, the vector sequence does not appear in the raw shotgun reads.

In cases other than the exception noted above, the vector sequences must be detected in order to be removed. However, identifying vector sequences at the 5'-end of raw shotgun reads with high accuracy is not a trivial task.

A simple way to mask vector sequences is as follows. In the first step, a Smith-Waterman alignment is performed between the raw shotgun reads and the vector sequence. When the alignment score exceeds a predetermined threshold, the detected alignments are masked as a vector. This

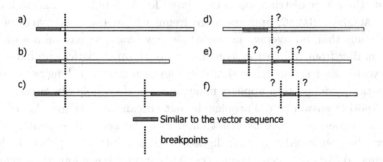

Fig. 8.24 Various types of vector trimming.

method is used in cross-match/phrap system developed by Phil Green [32]. One problem is that the base quality of the 5'-end of raw reads is extremely low, so that the threshold must have a very low value to ensure that all vector sequences are masked. Suppose that an alignment of 13 bp with 62% identity must be masked. It is obvious that nonvector sequences can easily be masked under such conditions; however, if a higher threshold to reduce false-positives is imposed, many vector sequences may not be masked, confusing the whole genome assembly. Clearly, there is no perfect solution or "de facto standard" for this problem.

One way of dealing with the problem is by adding heuristic rules. For example, the RAMEN assembler[2] has a three-tiered approach, in which a raw read is first searched by a BLAST-like algorithm. If the head part of the raw read is of high enough quality, the vector sequence is found; this represents the simplest case. Second, if the vector is not found by the fast BLAST-like search, the raw read is searched by the Smith-Waterman algorithm. This differs from the simple implementation mentioned above in that the threshold is dynamically determined according to the alignment. If the alignment is very close to the cloning site in the vector sequence and is also at the head part of the raw read, the alignment is likely to indicate that the vector has been found. However, if the alignment is far from the cloning site in the vector and is near the center of the raw read, it may only be found by chance. The threshold is determined for each Smith-Waterman alignment according to the positions of the alignment. There are cases when even these two approaches cannot detect the vector sequence in the raw read. Third, if the vector sequence is not discovered by this time, the head part of the raw read is reduced by 10 to 20 base pairs.

Another heuristic for the problem is implemented in ARACHNE2 [39]. First, the k-mer distribution is calculated for each library. If some k-mer, e.g., ACCGAGTTGAGAC, appears more frequently in the head parts of the raw reads than on average in the whole raw reads, it may be a sequence originating from the vector sequence. Relying on statistical probabilities, the vector sequence with low similarity can then be cut, as long as the same mutation/indel pattern appears many times in the whole raw reads.

Another answer to the problem is that "it can wait" (Figure 8.25). The vector sequences are usually not similar to the genomic sequence; thus, it can be concluded that the aligned parts of reads are unlikely to have originated from the vector sequences. This approach is adopted by genome

---

[2]RAMEN assembler is developed by Masahiro Kasahara and Shin Sasaki. It was used to determine the silkworm genome and the medaka genome.

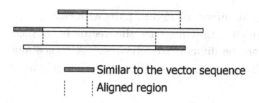

Fig. 8.25 Vector masking by alignment.

assemblers such as PCAP [38] and phrap [32]. Initially, this approach appears to work well even while maintaining simplicity; however, three points must be remembered. First, there is a trade-off between the accuracy of vector masking and the accuracy of repeat boundary detection, as will be described later. Second, raw reads are longer than trimmed reads and may cause a major increase in computation time. Third, sister clones, which is defined shortly, may confuse the strategy. After the transformation in the library construction step, the host *E. coli* must be cultured before planting in order to recover from cell-wall damage caused by electropolation. Although this step is accomplished in a short amount of time, a few bacteria may have proliferated before planting onto the plates. Because the bacteria can be planted at different places on the plates, the same clones can appear in the shotgun reads. These clones are called *sister clones*. It is obvious that whole raw reads of sister clones may be aligned well together, which violates the rule that "parts of vector sequences are unlikely to align well with other reads." It should be noted that carefully worked-out protocols can reduce the number of sister clones, so the proportion may be quite low in large-scale projects.

Even when all of these strategies are combined, however, difficult cases remain. Reconsider Figure 8.24. Figure 8.24d illustrates a raw read that contains part of the vector sequence in the center part, not in the head; thus, it cannot be determined what is in the head part of the raw read. Another problematic case involves multiple occurrences of vector sequences (Figure 8.24e). The usual approach to these cases is to simply discard them. Because they are usually inferior in quality to other raw reads, incorporating these cases into the assembly would not likely improve the results. Another case involves both a short and weak homology to the vector (Figure 8.24f). An alignment of 200 bp with 98% identity can easily be masked because such alignments never occur between vector sequences and the target genome as long as the target species is not evolutionarily close. Consider an alignment of 25 base pairs with 80% identity, which could be a genomic sequence that matches the vector by chance; intermediate cases may therefore be difficult to determine. A choice must be made by considering whether the subsequent algorithms are tolerant of the possible errors.

8.8.2 Quality Trimming

Even after vector masking, not all of the remaining parts of raw reads are useful for assembly. A typical raw read has a low quality area in the latter part (Figure 8.26). Because the quality of these bases is so low, nothing can be determined about them, and they are usually trimmed to reduce the amount of data, thus simplifying subsequent procedures. Three major algorithms are available for trimming raw reads based on their quality.

One way is to limit the expected number of erroneous bases in a read. A raw read usually contains a region of high quality; for example, the longest stretch of QV greater than 30 can be easily identified, as it is the core of the trimmed read. Based on the QV assigned to each base, the core is extended to both sides until the expected number of cumulative errors reaches a predetermined threshold. A variation of this algorithm limits the average error ratio of a trimmed read to a predetermined threshold. This approach has an advantage in that it provides a theoretical basis for the later stages of assembly algorithms. As the error ratio in a read or the cumulative number of erroneous bases is limited, it is easier to determine the width of a band in banded alignment.

However, the *window-trimming approach*, which is explained below, is often employed because of its practical fit to the problem. Due to various technical reasons, including secondary structure problems of Sanger fragments, inadequate elimination of fluorescent dyes, and nonuniform incorpo-

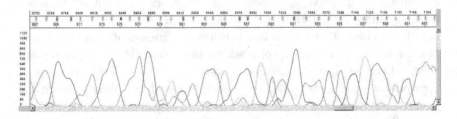

Fig. 8.26 The peaks are weak in the latter parts of raw reads. The shown read is from the medaka WGS project.

ration of ddNTPs, the peaks are often locally disturbed in electrophoretic analysis. As QVs depend on the evenness of peak spacing in a window of five peaks, QVs often drop down markedly within 5 bp around the disturbance. The approach in the previous paragraph tends to conservatively trim raw reads in the disturbed region, but many high quality bases are often left beyond the disturbance. The window-trimming approach scans through a raw read in a window of 10 to 20 bases. A window is defined to be "good" if the expected number of erroneous bases is less than some predetermined threshold, and raw reads are trimmed to the outermost good windows. Since the window trimming approach allows for disturbances in the center portion of reads, longer reads can be obtained, which leads to a smaller number of gaps in the assembly. Bases with a QV <9 are rarely useful for whole genome shotgun assembly [22].

The last approach again is that "it can wait." This approach does not trust QVs. As long as a sequence aligns well to other reads, it can be useful for the assembly; in our experience, however, low quality sequences rarely improve the assembly. The only exception is that a sequence of low quality may fill a gap between two contigs. Because the number of gaps between contigs is usually far less than the number of raw reads, a special process for gap filling may be appropriate for this purpose in terms of processing time.

When a small number of bases, e.g., less than 50 bp, remains after quality trimming, such a read is usually discarded, because short reads often complicate the assembly problem and have little potential to improve the assembly.

8.8.3 Contamination Removal

Although it depends on the target species, DNA sequences from sources other than the target nuclear genome often contaminate shotgun libraries. For example, when the DNA source of a shotgun library is a whole body of an adult fish, this library may also contain the genome sequences of aquatic microorganisms. Especially when the target species has a symbiotic lifestyle, the shotgun library may contain a huge amount of the genome sequences of other organisms. However, the most common contaminants are vectors and their host bacteria in the typical whole genome shotgun sequencing.

To eliminate this foreign DNA, a homology search is often performed against possible contaminants. At the very least, vector sequences and the genome sequences of their hosts must be included in the contaminant database to avoid misassembly and unnecessary computations. It is highly recommended that all sequences previously sequenced at the same place should be included in the database. Otherwise, contigs from other species that are occasionally more than 100 kb may greatly misguide later analyses. The mitochondrial sequence of the target species should be included in the database if it is known. A larger database is not always better because it is computationally costly, e.g., to BLAST tens of millions of shotgun reads against the entire GenBank database.

8.8.4 Overlap and Layout

Overlap detection and the subsequent layout of reads are the main components of a whole genome shotgun assembler. We have thus far discussed the overlap phase and the layout phase as separate entities, but both are tightly interconnected in terms of practical implementation. We will describe a wide variety of practical techniques used in the overlap and layout phases. Some can be integrated and some cannot, and the optimal combination of these techniques has not been determined.

8.8.4.1 Seed and Extend

To detect overlaps, it is essential to align two reads to determine if they actually overlap. However, examining all pairs of reads is too time-consuming, so the *seed and extend* strategy is often used.

The key idea behind this strategy is that the search space is restricted by shared *seeds*. A seed usually indicates a consecutive or nonconsecutive 8-mer to 32-mer. For example, if two reads have same subsequence CAGCAGCATGGGCATCAGTACGTAC, they share a seed of 26 consecutive bases. When a 20-mer (or greater) seed is shared between reads, it is probable that the reads originate from the same region in the genome (Figure 8.27). When the target genome is not repetitive, a 20-mer seed can identify a

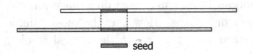

Fig. 8.27 A shared seed triggers dynamic programming alignment.

single location on the genome. Therefore, performing a pairwise alignment may be worthwhile. Furthermore, the space of dynamic alignment can be restricted to a narrow band along the diagonal of the seed (Figure 8.28).

Some pairs that actually overlap may not share a seed because of sequencing errors. The longer the size of a seed, the more often true overlaps are missed due to the absence of shared seeds. Several approaches have been proposed to remedy this flaw.

One approach is to simply decrease the seed size. A consecutive 28-mer seed may not be shared by some truly overlapping reads, while a consecutive 14-mer seed must be shared. If this is not the case, such overlaps can be discarded because the quality of these reads is so low that removing them will not degrade the assembly, unless by chance that low quality read can fill a gap between contigs of high quality reads. When the seed size is small, e.g., 8- to 14-mer, specificity may not be very high; in other words, even if two reads share a seed, in many cases they do not actually overlap in the genome. As shorter seeds lead to an exponential increase in computation time, other heuristics, such as two-hit heuristics, are usually added as in BLAST; two-hit heuristics require two distinct shared seeds to trigger dynamic programming alignment.

The second approach exploits transitive overlaps to recover missed overlaps, as illustrated in Figure 8.29. When read a overlaps with read b and read b overlaps with read c, missed overlap $a \to c$ can be inferred by transitivity (Case 1); similarly, $a \to b$ can be inferred in Case 2 of Figure 8.29.

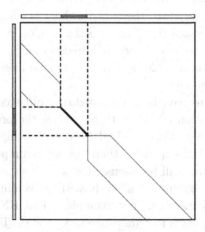

Fig. 8.28 The search space can be restricted to a narrow band.

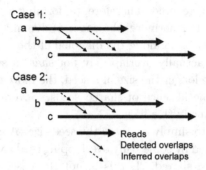

Fig. 8.29 Overlaps are inferred by transitivity.

The third approach once again is that "it can wait." Although abandoning the problem may degrade the quality of the assembly, this issue will be addressed later.

8.8.4.2 *Seeding*

The seed and extend strategy requires efficient seeding algorithms. One approach is to create a hash table that holds every occurrence of k-mers in the shotgun reads, as described in the previous chapter. For every read, reads with shared k-mers are retrieved from the hash table, after which they are aligned pairwise to determine whether they overlap. This scheme is rather simple and very fast; the only drawback is a huge memory requirement, which is an obstacle for assemblers that run on inexpensive machines. Without some compression, several hundreds of megabytes will be needed to store the entire hash table, and it may not be affordable for everyone assembling whole genomes.

Several algorithms have been proposed to avoid excessive memory consumption. One solution is simply to parallelize the process, utilizing the distributed hash table. One hundred PCs, each equipped with 2 GB of memory, are much less expensive than one supercomputer with 200 GB. Distributed algorithms will be discussed in a later subsection.

The following programs are all dedicated to avoiding the consumption of large amounts of memory. For example, ARACHNE sorts all 24-mer occurrences in 100 passes to reduce memory usage[3]. The concept of min-

---

[3] Considering symmetrical redundancy, ARACHNE actually sorts only half of the 24-mer occurrences.

imizer [84] is excellent to reduce memory usage by an order of magnitude at the expense of decreasing the sensitivity of overlap detection. It limits the positions at which seeds are sampled from reads while considering the surrounding nucleotide sequences. When the same nucleotide sequences are detected, the positions are limited in the exact same way.

8.8.4.3 Greedy Merging Approach

After detecting overlaps, the reads are laid out; while there are many strategies to compute the layout, most are based on the *greedy merging approach*. To explain the advantage of the greedy merging approach, we define *read flow graph* $G = (V, E)$, where V contains all double-stranded reads, and for every double-stranded read r_i, E contains at most one edge $r_i \to r_j$, where r_j is the read to which r_i overlaps. It is obvious by this definition that the read flow graph is a subgraph of the overlap graph; the only difference is that the number of outgoing edges from a vertex in the read flow graph is limited to one at most. To construct a read flow graph, one of the edges in the overlap graph is chosen for every vertex if it has multiple outgoing edges. How the edges are chosen depends on the strategy used; here, we initially use the greedy strategy to choose the overlap with the best alignment score.

We first illustrate how well the strategy works for small repetitive sequences, after which we show difficult cases. A typical example of small repetitive sequences is shown in Figure 8.30. Blocks A, B, C, and D represent unique sequences in the genome, while block R represents a repetitive sequence. The figure shows only parts of the genome; other parts, as well as the complementary sequences of the blocks, have been omitted to simplify the explanation. We assume R to be much shorter than the average length of shotgun reads, because most, but not all, repetitive sequences in

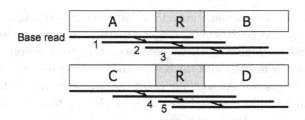

Fig. 8.30 Small repeat sequences are spanned by equal to or more than one reads.

genomes are relatively short. Below the blocks, the horizontal bars represent the shotgun reads that originate from the corresponding genomic region illustrated just above them. The base read has five overlapping reads, labeled 1-5. For the initial read flow graph, an outgoing arc with the best alignment score is selected for each vertex, thus, we have selected an arc from the base read to read 1. As an alignment score is generally proportional to the length of its alignment, the overlap with the longest overlap alignment is chosen for each read. In a similar way, the edges of read 1 → read 2, read 2 → read 3 are chosen. The correct layout, base read → read 1 → read 2 → read 3, is reconstructed in the read flow graph.

We may have committed an error if an edge from the base read to other reads, e.g., read 4 or 5 is chosen in the read flow graph. How is the correct next read chosen? As long as the target repetitive sequence is much shorter than the average length of the shotgun reads, it is likely that every occurrence of the repetitive sequence is completely spanned by at least one shotgun read. Once the spanning shotgun read is chosen, it contains unique sequences at both ends of the read, which helps avoid joining two irrelevant regions. Such spanning reads should always be exploited as long as the read flows are constructed by "walking little by little", i.e., the overlap with the best alignment score is chosen for each read.

Although it depends on the criteria used in quality trimming, the average length of shotgun reads is typically 600-800 bp, while most of the repetitive sequences are <300-400 bp in length. Although not all problems can be resolved in this way, small repetitive sequences are easily overcome by the greedy merging approach. The approach has been adopted by virtually all whole genome shotgun assemblers that employ the overlap-layout-consensus strategy.

8.8.4.4 Longer Repeat Sequence

We assumed in the previous subsection that the target repetitive sequences are relatively short. However, longer repetitive sequences also occur in genomes. For the longer repetitive sequences, the greedy merging approach fails because the overlap with the best alignment score does not always lead to the correct answer. Although in practice the correct overlap for a read is usually found within the overlaps that have the top five alignment scores, an overlap with the lower score may be correct in certain rare cases.

One example is shown in Figure 8.31. Blocks A, B, C, and D represent unique sequences as before. Block R represents a repetitive sequence in

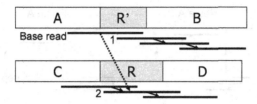

Fig. 8.31 A repeat sequence, R, and a truncated repeat sequence, R', which disturbs the read flow graph.

the genome, while block R' is a truncated R, which frequently appears in various genomes because incompletely retrotransposed elements often create this situation. The correct destination of the base read in the figure is read 1; however, the best overlapping read for the base read is read 2, which results in a contig with misjoins. One might assume that the branches in the read flow graph would reveal where the misjoins occur. In the figure, it is apparent that the branch at read 2 shows the conflict; thus, while it is true that the graph reveals the misjoins in some cases, it does not hold in all cases. Another sequence R", which is also truncated R, can be added to create an example in which there are misjoined flows with no branches in the initial read flow graph. Concrete examples are left for the reader to deduce independently, although an obvious example is one in which R" is truncated at the end opposite to that in R'.

8.8.4.5 *Iterative Improvements*

Most whole genome shotgun assemblers adopt the greedy merging approach as their basis, while several other strategies are added to avoid misjoins described in the previous subsection. One solution to the problem of longer repeat sequences is *iterative improvement*.

Assume that the initial read flow graph contains a misjoin caused by a medium-sized repetitive sequence (Figure 8.32, left). There are two choices for the branching read in the figure, which represents two distinct flows, route A and route B. The overlap to route A is initially chosen because it has the best alignment score. If a mate pair spans the branch, e.g., a read before the branch and a read on route B consisting of a mate pair, it is reasonable to replace the edge from the branching read to route A with the edge to route B (Figure 8.32, right). As long as the number of consistent mate pairs increases, the read flow graph is iteratively modified

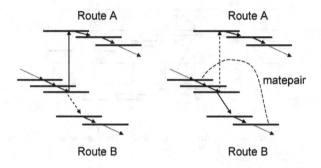

Fig. 8.32 The layout is iteratively improved by mate pairs.

accordingly. Assemblers like JAZZ[5, 93]/PCAP[38] use this strategy to improve the assembly.

8.8.4.6 *Accelerating Overlap Detection*

We previously explained the seed and extend strategy for overlap detection, in which a hashing scheme (or some other equivalent) narrows the possible overlapping pairs of shotgun reads to reduce the computation time. However, certain types of small repetitive sequences occur too frequently in the genome, especially when the genome of the target species is as complex as that of mammals or other vertebrates. For example, a repetitive sequence named *Alu*, which is about 300 bp long, occupies approximately 5% of the human genome [20]. The number of shotgun reads that contain the *Alu* sequence can be calculated assuming 8x coverage, which is a typical value for whole genome shotgun sequencing. Assuming the average length of shotgun reads is 700 bp and the genome size of humans is 3 Gb, the number of reads is $\frac{3\times 10^9 \times 8}{700} \times 0.05 \sim 1.7 \times 10^6$. Although several mutations occur among the copies that distinguish them in some types of hashing schemes, it is apparent that the usual hashing schemes that exploit short oligomers may not be useful for eliminating the search space of overlap detection. Because the number of edges in the overlap graph may quadratically increase as the occurrence of repetitive sequences increases, short repetitive elements like short interspersed nuclear elements (SINEs) must be masked during the overlap detection phase; otherwise, too many overlaps fill up the main memory or external storage, which leads to an unacceptable increase in computation time.

One solution is to simply exclude frequent short oligomers

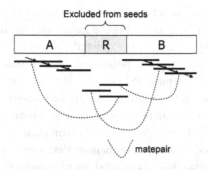

Fig. 8.33 The reads filling the gap are recruited by mate reads on the unique regions.

for seeds; fragment assemblers like ARACHNE[7]/RePS[103]/JAZZ[5, 93]/PCAP[38]/PHUSION[64] use this strategy in an explicit or implicit manner. The 16- to 32-mers occurring over 50 to 300 times in given shotgun reads are usually excluded from the seeds. Even if such seeds are shared between distinct shotgun reads, further alignments by dynamic programming are not triggered, which saves a great deal of space and time and consequently allows the computation of the overlap graph to be accelerated by at least one order of magnitude.

Will the simple exclusion of seeds of short oligomers that occur too frequently degrade the assembly? This would be the case if both the computation time and the space requirements were ignored, but both are limited in practice. It is necessary to mask out such sequences from seeding, even though the masking makes it impossible to create a correct overlap graph around the frequent repeat sequences.

Provided that the seeds are not located in the repeat intensive region on the genome, the problem can be easily remedied. Even when some reads are replete with repetitive sequences that appear too frequently in the genome, the mate reads may not be. Because most of the problematic repetitive sequences tend to be interspersed, even a read that consists solely of repetitive sequences can be placed confidently when its mate read contains unique sequences that can be placed unambiguously in the assembly (Figure 8.33). These repetitive reads are locally assembled to patch gaps. Note that this strategy may remedy most problematic cases, though not all of them. When both ends of a mate pair consist solely of frequent repetitive sequences, they cannot be reliably placed in the assembly.

This extensive masking is aimed at speeding up the computation of overlaps, not at masking all repetitive sequences to avoid misassembly. Repetitive sequences that appear several times in the genome may not be masked. With 8x shotgun sequencing, it is expected that unique sequences are covered 8 times on average, while repetitive sequences that occur twice in the genome are covered 16 times on average. However, due to the stochastic nature of DNA fragment sampling, unique sequences may be covered 16 times, while repetitive sequences that occur twice may be covered only 8 times at a given probability. Assuming Poisson statistics, we conclude that unique sequences and repetitive sequences that occur twice cannot be detected reliably using the frequency of short oligomers in the shotgun reads.

8.8.4.7 Repeat Sequence Detection

One of the biggest challenges facing whole genome shotgun assemblers is to avoid misassembly caused by repetitive sequences in the genome. If repetitive sequences could be reliably detected, assembly algorithms would be straightforward. In reality, however, several different strategies must be combined to predict repetitive sequences more accurately. Various algorithms for detecting repeat sequences are described below.

One algorithm for this purpose is the "repeat database" strategy, which performs a homology search against repeat sequence databases. For example, repetitive sequences in the human genome are well-annotated so that most are recorded in the database. Querying shotgun reads against the database provides definitive evidence for repetitive sequences. The Celera assembler [66, 67] used this strategy to reconstruct the human genome using the whole shotgun method. Although it achieves the highest specificity among repeat sequence detection algorithms, it is obvious that this strategy may not always be applicable to other sequencing projects. This strategy requires well-annotated repeat sequence databases that cover all repetitive sequences in the target genome. While the human genome has been studied extensively, the repetitive sequences of many other species are not well known.

Another algorithm involves masking short oligomers overrepresented in the shotgun reads, as described in the previous subsection. The advantage in this strategy is that a repeat database does not need to be created beforehand. A threshold T is established, and short oligomers that occur more than T times are masked as repeats. A repeat sequence occurring 100 times in the genome is expected to appear in the shotgun reads 100 times more often than unique sequences, and more than 800 times in the shotgun reads using 8x coverage. In contrast, short oligomers (e.g., 20 bp) in the unique region of the genome seldom appear more than 30 to 50 times in the shotgun reads according to the shotgun coverage. Thus, T is usually set to a value ranging from 30 to 100. However, this is not a perfect solution.

First, due to the stochastic nature of fragment sampling, a repetitive sequence with a relatively low number of occurrences may not be reliably distinguished from unique sequences. The accuracy of repeat detection can be estimated assuming an ideal Poisson distribution. The condition assumed here is the same as the one described in the section on Lander-Waterman statistics; the effective sequence coverage is λ. The probability $P_1(c)$ of having coverage c at a base in a unique region of the genome is

$$P_1(c) = \frac{e^{-\lambda}\lambda^c}{c!}$$

The probability $P_2(c)$ of having coverage c at a base in a genomic subsequence that occurs twice is only

$$P_2(c) = \frac{e^{-2\lambda}(2\lambda)^c}{c!}$$

When the ratio of false-positives is limited to 0.001, which is not a very low value, T must satisfy the following condition:

$$\sum_{i=T+1}^{\infty} P_1(i) < 0.001$$

At the typical shotgun coverage, 8x, or $\lambda = 8$, $T = 19$ is required to meet this condition. However, the sensitivity S of detecting the repetitive sequence that occurs twice is

$$S = 1 - \sum_{i=0}^{T-1} P_2(i) = 0.258$$

Thus, more than half of the repetitive sequences would not be detected.

Second, although shotgun reads are sampled uniformly at random under ideal conditions, in reality, some types of repetitive sequences tend to be underrepresented in shotgun reads. For example, some types of sequences have hazardous effects on the host bacterium, so that they never or seldom appear in the shotgun library. As another example, AT-rich, GC-rich, and inverted repeat regions are difficult to sequence using the typical protocols for massive sequencing. As a result, short oligomers in such regions appear less often, making it impossible to distinguish them from unique oligomers.

For example, the medaka genome contains several copies of *tol2* ([51]), which is a known repetitive sequence with an direct inverted repeat, whereas the number of shotgun reads containing the *tol2* direct inverted repeat was far less than expected from the wet experiment. (A. Koga, Nagoya Univ. personal communication)

Another algorithm to detect repeat sequences is the *repeat boundary* approach, which detects the boundaries between unique regions and repetitive regions. As long as the boundaries are all covered by the shotgun reads, two conflicting overlaps as illustrated in Figure 8.19 must be observed for each boundary. These boundaries are called *repeat boundaries*. When creating contigs, merging can be stopped at the repeat boundaries to avoid misassembly caused by repetitive sequences. If all repeat boundaries are well-covered by the shotgun reads and are completely detected as expected, the number of misassemblies will dramatically decrease. However, in reality, the presence of both sequencing errors and polymorphisms complicate the problem.

An example is shown in Figure 8.34, in which the reads are truncated to fit on the page; the actual reads are much longer than illustrated and contain many more insertions/deletions and mismatches. There are at least three possible scenarios for the three shotgun reads. The first is that one deletion and two mismatches in the alignment between read B and read C arise from mere sequencing errors. The second is that some discrepancies are due to repetitive sequences, while others arise from sequencing errors. Two sequences, R and R', differ by several base pairs. The third is that some of the discrepancies are due to polymorphisms. When sequencing a bacterial genome, the source DNA is usually extracted from the bacterial population. Are the genomes exactly the same? Generally speaking, the answer is no. There may be indels and mutations, or even rearrangements, though they are rarely observed in bacterial sequencing projects. When sequencing a diploid genome, it must have two different haplotypes, unless the DNA library is from an inbred strain. There are polymorphisms between the two haplotypes, which may lead to discrepancies. In the worst-case scenario, sequencing errors, repetitive sequences, and polymorphisms can all combine to yield slightly different but apparently overlapping shotgun reads. A sequencing error ratio of 5% is common at the tails of shotgun reads, whereas repetitive sequences with over 95% identity (sometimes even 100%) can be seen in the genome. In general, the polymorphism ratio between two haplotypes ranges from 0 to 30%, which is highly dependent on the target species or strain. Therefore, simply setting a threshold for alignment identity will not solve this problem.

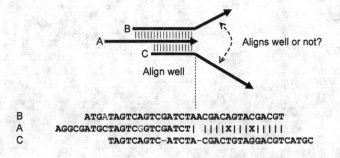

Fig. 8.34 Is this a repeat boundary?

8.8.4.8 Error Correction

To simplify the repeat boundary detection problem, an *error correction* is performed in several assemblers before both the overlap detection and repeat detection. The key idea is that sequencing errors tend to occur independently. At 8x shotgun coverage, most parts of the genome are covered by more than two shotgun reads. As the sequencing errors are "outvoted" in the multiple alignment of the overlapping reads, the correct bases are easily estimated (Figure 8.16, 3). In the case of a tie, bases in a column are usually weighted by their QVs. Even though the error correction process sometimes introduces several errors into the shotgun reads, the number of sequencing errors usually decreases by an order of magnitude, greatly simplifying repeat detection. Repetitive sequences with slight discrepancies and polymorphisms may collapse if a particular base of either a repetitive sequence or one haplotype is covered by only one singleton read. Bear in mind that some types of sequencing errors are not independent and easily reproduced; thus, they are not easily corrected by simple multiple alignment.

8.8.4.9 Repeat Separation

If the source DNA of the shotgun libraries is derived from an inbred strain or bacterium that proliferates rapidly, polymorphisms can practically be ignored because they are rarely observed (e.g., $< 10^{-5}$). In such cases, the

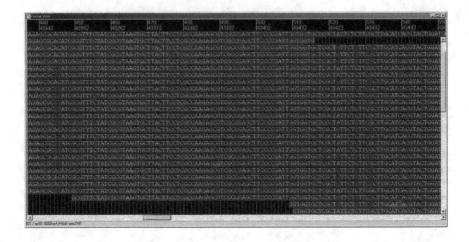

Fig. 8.35 Multiple columns support two distinct sequences. The raw shotgun reads can be downloaded from *Cyanidioschyzon merolae* project page (http://merolae.biol.s.u-tokyo.ac.jp/). The multiple alignment is generated by RAMEN assembler.

repeat separation strategy can be easily applied to improve the continuity and accuracy of the assembly. Two (or more) copies of a repetitive sequence often differ by at least several percent because the bases have mutated during the course of evolution. If the mutations can be detected among the copies, the collapsed repeat sequences (Figure 8.35) will be separated into distinct copies.

However, not all sequencing errors are eliminated by this phase. Multiple instances of sequencing errors sometimes by chance display the same nucleotides, which pass through the error correction phase. To reduce the effect of sequencing errors, the QVs of the conflicting bases within a row are checked for reliability. For example, if A,C, A, C and C are in a row and all have a QV > 40, this would strongly suggest that they originated from different regions in the genome.

Note that this reasoning is based on the assumption that polymorphisms are rarely observed and can thus be ignored. If there are two distinct haplotypes that have diverged at a rate of a few percent, they may exhibit a multiple alignment with conflicting bases as though there were multiple copies of a repeat sequence.The major difference is that the local sequence coverage for one haplotype is half of the non-diverged region in the latter case, while it is doubled in the former case. As described previously, the

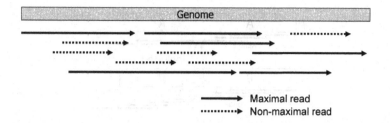

Fig. 8.36 Maximal reads cover the genome. Non-maximal reads can be temporarily set aside.

local sequence coverage depends on stochastic processes and is therefore unreliable.

Even when the polymorphism level is so low that polymorphisms can be ignored in most cases, some polymorphisms occur mainly due to replication errors. Repeat separation algorithms will separate these polymorphisms into different contigs, which may lead to a break in a contig. To remedy this situation, gap filling should be performed (Figure 8.33).

8.8.4.10 *Maximal Read Heuristics*

Constructing the overlap graph for given shotgun reads requires a considerable amount of time. *Maximal read heuristics* eliminate non-maximal shotgun reads, which are entirely within other shotgun reads, to reduce time and space consumption during the construction of the graph. The idea is that non-maximal shotgun reads can be put aside during the process of constructing the overlap graph and during the early layout phase. Assemblers such as ARACHNE employ this idea. Non-maximal reads are set aside during the construction of the overlap graph until the tiling path of each contig is computed (Figure 8.36). Non-maximal reads that can be unambiguously mapped are then placed on the tiling paths. At 8x shotgun coverage, the number of maximal reads is typically half of all shotgun reads.

The concept of maximal read heuristics is relatively simple with regard to unique regions of the genome, although it may complicate construction of the overlap graph because maximal reads may overlap with non-maximal reads in the repetitive regions of the genome (Figure 8.37). This figure shows a read at the boundary between B and R, called the "boundary read," which overlaps with a non-maximal read in R. The non-maximal read is termed the "child," while the maximal read that contains the child read is termed the "parent," as illustrated in the figure. In the figure, the

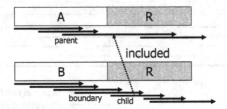

Fig. 8.37 Maximal read heuristics complicates the overlap graph construction.

parent read that contains the child read has a unique A flanking at one end, which makes it impossible for the boundary read to overlap with the parent. This does not occur in unique regions of the genome.

8.8.4.11 Paired Pair Heuristics

Some types of repetitive sequences, including heterochromatic repeats, are very difficult to assemble. In particular, long tandem repetitive sequences in which a single unit is composed of a short sequence of up to 100 bp are extremely difficult to reconstruct. In contrast, other types of repetitive sequences, including SINEs and LINEs (Long Interspersed Nucleotide Elements), are interspersed in the genome. It is generally difficult to check whether two shotgun reads that contain the repetitive region of the genome actually overlap. However, if their mate reads overlap, it is highly probable that the two reads actually overlap in the genome, except when the mate reads are also involved in the repetitive regions of the genome (Figure 8.38). Assuming that the probability of encountering exceptional cases is low, at least for interspersed repetitive elements, two overlapping mate pairs are merged by priority in the early stage of the layout phase; this is called *paired pair* or *sister edge* heuristics.

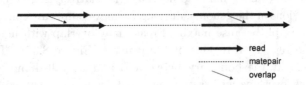

Fig. 8.38 Paired mate pairs overlap.

8.8.4.12 Parallelization

Although it depends on both the algorithm chosen and the computer used, it is not unusual that several days to a whole month are needed to construct the overlap graph for tens of millions of shotgun reads. To reduce the turnaround time, many researchers have attempted to construct overlap graphs in parallel.

A long turnaround time leads to inefficiency in genome assembly. For example, one may hope to minimize the number of shotgun reads to save money, but before actually assembling the shotgun reads, predicting the actual size of scaffolds is extremely difficult. If the whole assembly process takes only 1 or 2 days, all currently available shotgun reads can be assembled to determine if we can stop adding reads. When a sufficiently large number of good scaffolds is attained, no additional shotgun reads are needed. In this case, target-specific sequencing is more appropriate than adding shotgun reads. However, if the whole assembly takes 3 months, for example, researchers may fail to switch their efforts from obtaining shotgun reads to another procedure. As explained in the section on Lander-Waterman statistics, the contiguity of the assembly increases exponentially as the coverage increases. The turnaround time should be as short as possible, especially in the late phase of the whole genome shotgun sequence assembly.

Another concern is that technical problems may arise in the later phases of a project. Assembling a whole genome is not yet a routine task, at least for a genome of more than several hundred mega bases. For example, many projects have had to deal with abundant polymorphism, which were not foreseen prior to assembling the shotgun reads. Other projects have suffered from extremely biased shotgun clones. When an unpredicted problem is detected, it is often recognized in the later phases of the project. Creating new libraries, modifying the assembly algorithms, and other efforts must be performed to ameliorate the problem, although a long turnaround time may often deter prompt and adequate solutions.

It is fairly evident that a simple BLAST search against large databases can be easily computed in parallel. If access to multiple CPUs is available, the databases can be divided into several smaller ones, and each CPU can then process the search against part of a large database. In contrast, the construction of an overlap graph is not easily divided into smaller tasks.

One of the naive but realistic parallelization procedures involves dividing the shotgun reads into smaller parts. For example, assume that shotgun reads are divided into 100 parts, e.g., $S_1, S_2, \ldots, S_{100}$. To enumerate all overlaps, the subset of overlaps is computed for each pair of $S_1, S_2, \ldots, S_{100}$. For example, the first subset contains only overlaps from r_i to r_j, where $r_i \in S_1$ and $r_j \in S_2$. As a result, the whole task is divided into $100 \times (100+1)/2 = 5050$ subtasks. Any group of tasks is computed independently, so that parallelization can be utilized. The complete overlap graph is constructed by simply merging individual graphs together. The major drawback of this algorithm is that the total computation time increases quadratically as the number of partitions grows. Most variations of overlap detection algorithms exploit some hashing scheme to achieve expected $O(1)$ time to find shared substrings, while the parallel algorithm described above requires expected $O(N)$ time in total, where N is the number of the subtasks, which grows quadratically.

The assembler PHUSION [64] adopts a unique clustering algorithm that divides given shotgun reads into clusters according to shared substrings. The key idea is that if two shotgun reads share many substrings, they can be assembled into one contig. The minimum number of shared substrings, T, and the size of substrings, k, are set before assembling; every pair of shotgun reads is then clustered into one when they share more than T k-mer substrings. For example, two reads are clustered when they share ATGGGTAC, AACGAACT, and GATGCACC, where $T = 3$ and $k = 8$, although the values of T and k are much higher in actual projects. After clustering, every cluster can be assembled separately in parallel. If the sizes of the clusters are too large, the parameters are altered to decrease the sizes to fit into the physical memory of the machines. While this concept is easy to understand, the problem is that the accuracy of the assembly relies on the accuracy of the clustering. As experimental errors, including sequencing errors, tend to occur more frequently in specific regions of the genome, shared k-mer clustering may potentially limit the ability to merge shotgun reads from such regions. The PHUSION assembler iteratively improves the generated contigs by using both mate-pair information and homology between the contigs to recover links between the clusters. However, the amount of heuristics used makes it difficult to evaluate and predict the theoretical result of the assembly.

8.8.4.13 Eliminating Chimeric Reads

As described earlier, some shotgun reads contain chimeric fragments as a result of experimental limitations. A *chimeric read* usually consists of two

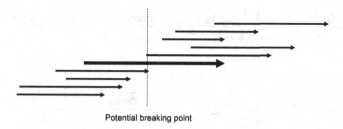

Fig. 8.39 A read with a potential breaking point. No overlapping reads align beyond this point.

different parts of the target genome and may connect two different parts of the genome into one contig, which will confuse the assembly. Therefore, it is essential to eliminate chimeric reads, or at least acknowledge them for later algorithms. To detect chimeric reads, potential breaking points are calculated for each read. A potential breaking point is a point in a shotgun read at which no overlapping reads align (Figure 8.39). Even in a chimeric read, the breaking point may be covered by alignments against overlapping reads, although they do not extend more than several bases beyond the breaking point because they only occur by chance.

This heuristic eliminates almost all chimeric reads, although it also eliminates non-chimeric reads that have potential breaking points by chance. To avoid eliminating useful reads, the reads with breaking points are usually examined further to determine whether their breaking points are covered by mate pairs. The presence of covering mate pairs ensures the correctness of the assembly over the potential breaking points.

8.8.5 Scaffolding

In the scaffolding phase, the contigs are connected by mate-pair information suggesting the relative order and orientation of the contigs. At first glance, this appears straightforward, but chimeric clones and human error complicate the scaffolding process. Because a chimeric clone has two distinct DNA fragments from different parts of the genome, the mate pair may connect two irrelevant contigs that are actually far apart in the genome. Although the percentage of chimeric clones is less than 0.1%, we cannot rely on the accuracy of all mate pairs.

One well-known heuristic that is used to avoid this problem is to discard the weak links, each of which consists of a single mate pair. With 8x

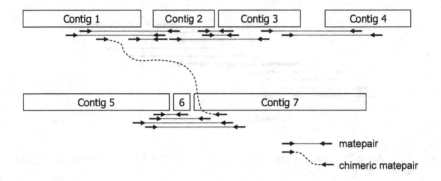

Fig. 8.40 7 contigs and 13 mate pairs are shown. The chimeric mate pair connects two irrelevant scaffolds.

sequence coverage, it is likely that two adjacent contigs will have more than one mate pair, while chimeric mate pairs are unlikely to connect the same pair of contigs twice. However, discarding all weak links greatly limits the ability to extend scaffolds. Using a careful experimental protocol, the percentage of chimeric clones is usually found to be far less than 0.1%, which indicates that most weak links are trustworthy and should not simply be discarded.

One common approach to the chimeric mate pairs is the *greedy scaffolding* strategy, in which more reliable links are used earlier to throw out the chimeric weak links that often cause conflicts. To demonstrate that it works well for actual shotgun reads, a typical case is shown in Figure 8.40, which shows seven contigs and 13 mate pairs. One of the mate pairs depicted on the dotted line is chimeric. The horizontal positions of the shotgun reads correspond to the actual positions on the contigs, suggesting their relative positions. The link between contig 5 and contig 6, and the link between contig 1 and contig 7, are the only weak links in the figure. Each link consists of a single mate pair, although the former link cannot be adopted without resolving a conflict with other reliable links, while the latter does not cause a conflict. For example, the mate pair between contig 1 and contig 7 suggests that both contig 3 and contig 7 coexist in the same region on the genome, which could never happen. In contrast, the weak link between 5 and 6 suggests that contig 6 fits in the gap between contig 5 and contig 7. In most of these cases, chimeric mate pairs connect two completely different parts of the genome, and thus can be detected by identifying conflicts with other reliable links.

The greedy scaffolding algorithm can be summarized as follows. First, each contig forms a scaffold. Second, links are picked up one by one in decreasing order of reliability, which is defined shortly. The chosen links are examined to determine if they can be adopted without causing a conflict with existing scaffolds. If they cannot, the two scaffolds are connected to form a merged scaffold according to the link; otherwise, the link is simply discarded as being chimeric.

The definition of reliability varies according to different assemblers, although common criteria include the number, lengths, and types of mate pairs in the link. Generally speaking, short plasmids are rarely chimeric, while longer clones are more frequently chimeric. Chimeric mate pairs are often attributable to human error, such as providing DNA sequencers with incorrect plates, which often results in incorrect mate pairs in the shotgun reads. In this case, chimeric mate pairs tend to appear in a rather systematic fashion, so they can be easily detected if the reason for their existence can be determined. However, some of chimeric mate pairs are often unable to explain.

8.8.5.1 *How Many Mate Pairs Are Needed for Scaffolding?*

An abundance of mate pairs is essential for scaffolding to be successful; otherwise, a large number of unrelated scaffolds would be presented. *Clone coverage* is defined as the average number of clones that cover an arbitrary position in the genome. The approximate distance between ends of the mate pair is called its *clone length*. The clone coverage is calculated as the total clone length divided by the estimated genome size.

A larger clone coverage in each cloning vector type is likely to increase the possibility of extending scaffolds. In reality, however, due to budget constraints, it is not practical to generate too many mate pairs; this necessitates a method for estimating the number of mate pairs required for successful scaffolding. This analysis requires precise and accurate information on the repetitive sequences in the genome of interest, which allows one to plan the collection of mate pairs that are able to sandwich repetitive sequences. However, prior to sequencing the genome, a nearly complete picture of repetitive sequences is generally not available. Such databases require complete genome sequencing. Metaphorically speaking, we are attempting to solve a jigsaw puzzle with many pieces from highly homogeneous regions without having seen the original picture.

Despite these difficulties, we can derive a rough estimate from previous genome sequencing projects. Table 8.2 presents the clone coverages of mate pairs used in individual projects that adopted the WGS sequencing approach. In Table 8.2, figures are cited from original papers [5, 25, 22, 40, 83, 21], although care should be taken in interpreting and comparing these figures because they are not calculated according to standardized criteria. For example, the average clone length is not always displayed, and more precise information such as the clone length distribution with the standard deviation is not available. For example, the thresholds of nucleotide quality values are not necessarily standardized, and there are distinct procedures for removing cloning vector sequences from DNA fragments.

Despite these issues, Table 8.2 provides an idea of how many mate pairs should be collected for sequencing large-scale vertebrate genomes. Note that several types and lengths of cloning vectors have been used, and their clone coverages are similar to some extent.

8.8.5.2 Iterative Improvements

Even if all of the algorithms described above are implemented carefully, some global/local misassemblies will remain, unless the target genome has a simple structure. These misassemblies arise mainly from human error, insufficient information provided by given shotgun reads, inaccurate detection of repetitive sequences, and polymorphisms between haplotypes. Greedy scaffolding often yields correct contig layouts, but it also has the drawback that when misassemblies occur in the contig-generation step or in the early phase of scaffolding, they will never be corrected. Since the accuracy of repeat sequence detection can never be 100% due to the stochastic nature of fragment sampling, and because other errors cannot be eliminated completely, an error-correction algorithm for misassembled contigs/scaffolds should be introduced. It is natural for misassembled contigs to have conflicting outgoing mate pairs. Therefore, the scaffolds often become shorter because of the misassembled contigs.

Most of the initial scaffolding algorithms proposed so far focus on local rather than global integrity, as shorter mate pairs are connected by priority. As a result, a misassembled contig often connects two irrelevant scaffolds into one, as shown in Figure 8.41. Provided that long-range linkages, such as fosmid or BAC end sequences, are available, signs of misassembly can be detected by mapping these end sequences onto the assembly; multiple instances of a long mate pair that conflicts with the existing scaffold will connect the same pair of scaffolds. This can be detected to fix the problem;

Table 8.2 Whole genome shotgun sequencing projects and their libraries. "-" signifies that the value is not mentioned in the paper. Note that *Rattus norvegicus* project adopted the hybrid sequencing approach, not the pure WGS approach.

Species	Year	Genome Size (Mbp)	N50 scaffold (Mbp)	N50 contig (kbp)	Library			
					Vector Type	Mean Length	Sequence Coverage	Clone Coverage
Fugu rubripes	2002	380	125kb	-	plasmid	2.0kb	2.26	3.33
					plasmid	2.0kb	0.44	0.64
					plasmid	2.0kb	0.21	0.35
					plasmid	1.9kb	0.07	0.10
					plasmid	2.1kb	1.38	2.39
					plasmid	2.0kb	0.82	1.15
					plasmid	5.4kb	0.32	1.18
					cosmid	39kb	0.05	1.65
					BAC	68kb	0.04	2.17
					Total		5.60	12.96
Ciona intestinalis	2002	160	190kb	-	plasmid	2.8kb	0.97	2.09
					plasmid	1.8kb	0.21	0.28
					plasmid	6.2kb	0.31	1.42
					plasmid	2.8kb	6.45	13.36
					plasmid	2.8kb	0.28	0.63
					cosmid	36.5kb	0.13	3.30
					BAC	120kb	0.03	4.11
					BAC	120kb	0.03	2.38
					Total		8.40	27.57
Mus musculus	2002	2500	16.9Mb	24.8	plasmid	2kb	0.71	1.20
					plasmid	4kb	5.89	17.70
					plasmid	6kb	0.22	1.00
					plasmid	10kb	0.52	4.30
					fosmid	40kb	0.26	9.30
					BAC	150-200kb	0.09	13.70
					plasmid	other	0.01	0.02
					Total		7.70	47.22
Rattus norvegicus	2004	2750	5301kb	38	plasmid	2-4kb	1.80	3.70
					plasmid	4.5-7.5kb	0.87	2.96
					plasmid	10kb	1.50	11.63
					plasmid	50kb	0.25	9.47
					BAC	150-250kb	0.07	9.26
					Total WGS		4.40	37.00
					BAC skims	2-5kb	2.90	4.80
					Total		7.30	41.80
Tetraodon nigroviridis	2004	340	984kb	16	plasmid	2-5kb	3.20	13.30
					plasmid	2-8kb	3.10	20.50
					plasmid	1.5-3kb	1.70	4.60
					BAC	100-160kb	<1	9.80
					BAC	120-180kb	<1	8.60
					Total		7.90	56.80
Drosophila pseudoobscura	2005	156	996kb	51.9	plasmid	2.7kb	2.06	4.06
					plasmid	3.4kb	5.73	13.60
					plasmid	6.3kb	2.73	14.67
					fosmid	40kb	0.12	3.27
					BAC	130kb	0.04	3.26
					Total		10.67	38.85

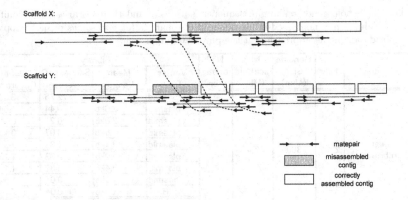

Fig. 8.41 Conflicting mate pairs span two scaffolds.

the two scaffolds can be broken and rejoined to satisfy more constraints imposed by the mate pairs. ARACHNE2 [39] achieved significant scaffolding accuracy using this iterative "breaking and rejoining" strategy.

8.8.6 Consensus

An assembler has to output a representative sequence of the genome in the final phase. As the ordered multiple shotgun reads generated in the layout phase are not suitable for later analysis, an assembler performs multiple alignments of the piled reads to create a consensus sequence. In [67], Myers describes this as follows, 'The quality of the trimmed sequence in Celera's data is so high that a simple shift-and-evaluate algorithm that we call "abacus" suffices to compute the optimal consensus-measure sequence.' As long as the polymorphism level is low and the shotgun reads are trimmed to retain only high-quality bases, building consensus sequences is a trivial task.

Iterative improvement heuristics, such as described in [4], change only a few base pairs per 100 kb after the Master-Slave alignment described in the previous section. However, the low-quality region should be treated using the traditional time-consuming multiple alignment algorithm because some regions are systematically of poor quality. For example, a long stretch of poly-"A" in the genome causes slippage of the polymerase, which decreases the base quality of the subsequent region.

The consensus bases are usually determined by majority decision, although majority decision sometimes fails to derive a correct consensus base.

For example, if there are three votes for "A" (QV=7, 9, and 11, respectively) and two votes for "T" (QV=20 and 25, respectively), which is the correct consensus base? Recall that the Quality Values of two shotgun reads may be highly correlated. Even if there are three votes for "A", their Quality Values are relatively low, so that systematic disruption is suspected. Since such disruption often depends on the upstream region of the shotgun reads, shotgun reads obtained from the opposite strand usually provide high-quality alternatives. Different sequencing protocols using different chemistry, dyes, polymerases, temperatures, or reagents may also allow us to obtain high-quality bases for such problematic regions. Therefore, several assemblers, including phrap, assume that the Quality Values of the bases from different strands are independent, while the Quality Values of the bases from the same strand are strongly correlated.

8.9 Quality Assessment

Once the assembly is completed, quality assessment is important to verify its correctness, which is absolutely necessary to ensure its quality.

N50 value

The length distribution of the output scaffolds is the most fundamental indicator of the quality of an assembly. Short scaffolds require additional shotgun reads, whereas long scaffolds can provide a good basis for later analysis. Although a cumulative plot of length distribution leads to complete understanding of the length distribution, a single representative value, which is called the *N50 scaffold length*, is often used to show the overall distribution of scaffold lengths. Using the average length of the scaffolds is one possible choice, although this tends to underestimate the quality of the assembly because short contigs that originate from organisms other than the target sometimes push the average length down considerably. By contrast, the N50 scaffold length, which is defined as the minimum length of scaffolds that span half of the nucleotides, is robust against added short contigs that arise from contamination, because small contigs have minimal effect on the N50 value. The *N50 contig length* is defined in a similar way. Note that N50 contig/scaffold lengths can easily be increased by misassemblies, so that these values must be evaluated together with other quality measures.

Comparison with Reference BACs

If some reference BACs, which are finished manually, are available, comparison with the reference BACs will show the actual assembly quality of those regions. Since the finished BACs are considered accurate at the base level (note that the degree of completion differs from BAC to BAC, so that their accuracy may differ) and they represent part of the genome, complete quality assessments can be done in those regions. Alignments between the scaffolds and reference BACs reveal the continuity and coverage of the sequence, and the accuracy of the contig ordering and orientation, gap size estimation, and each base. This is the only method capable of evaluating base level accuracy, although the finished BACs might be biased towards regions that are easy to sequence because BACs that contain problematic regions of the genome are often difficult to finish.

Comparison with Genetic/Physical Maps

The long-range integrity of the assembly can be measured by aligning genetic markers against the scaffolds, if the markers are available. It is better to have as few conflicts as possible, although the reliability of the genetic map depends on both the protocol and the material used and some errors may be observed in the map. Therefore, the number of conflicts may not be zero even if the scaffolds are completely correct. The same discussion applies to physical maps.

Comparison with Reference ESTs/cDNAs

Since ESTs and cDNAs are transcribed from the genome, ESTs and cDNAs must appear on the assembly. By aligning ESTs/cDNAs against the assembly, the quality of the assembly can be estimated.

The EST/cDNA coverage, which is defined by the proportion of these sequences covered by the assembly (i.e., covered by BLAST alignments), can serve as a quality measure of cloning bias. If the assembly does not cover a significant proportion of the ESTs/cDNAs, this suggests that the shotgun library is biased or that they are too fragmented due to misassembly. ESTs/cDNAs in the scaffolds may be fragmented for several reasons. Stochastic gaps in the assembly may split alignments, and smaller alignments must be counted as the coverage to measure the bias in the shotgun library. Note that coverage may be overestimated due to the presence of repetitive sequences in ESTs/cDNAs.

In turn, under other criteria that exclude smaller alignments, the local integrity of the assembly can be estimated. As the order of exons must be preserved between ESTs/cDNAs and the genome, even if splicing is considered, spliced alignments between ESTs/cDNAs and the genome suggest the overall accuracy of the local integrity (i.e., the accuracy of contig ordering and orientation).

Mate pair constraints

A pair of reads that originate from the same clone must be placed on the assembly, satisfying the following conditions in the ideal case.

(1) The two reads must face each other because they represent both ends of the same clone and they are from the opposite strands of the clone.
(2) The distance between the pair must be a biologically reasonable distance depending on the type of vector.

These conditions are called *mate pair constraints*. The proportion of satisfied mate pair constraints suggests the overall accuracy of contig ordering, contig orientation, and contig construction.

8.10 Past and Future

Various approaches for whole-genome sequencing have been proposed in the past few decades. Although the whole-genome shotgun method is now regarded as the standard technique for whole-genome sequencing, many doubted its applicability to large genomes, such as those of mammals [34, 33]. When the International Human Genome Sequencing Consortium launched the project to sequence the human genome, few were confident that the whole-genome shotgun method worked well, so they adopted the *BAC-by-BAC approach*.

In the BAC-by-BAC approach, the genome is first sheared and cloned into BACs. Using physical maps, the minimum number of BACs covering the genome is selected. The selected BACs are then shotgun sequenced individually, and then combined to reconstruct the whole genome sequence. The major advantage of this method is that it is less affected by the presence of repetitive sequences. Even if the target genome contains millions of copies of transposable elements, each BAC may only contain a few copies simply because each BAC is usually orders of magnitude smaller than the whole

Read: ACCGTGACT
ACC
CCG
CGT
GTG
TGA
GAC
ACT

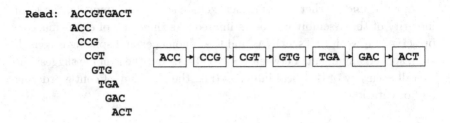

Fig. 8.42 A simple example of a de Bruijn graph.

genome. A smaller number of repetitive sequences allows the use of more stringent algorithms, facilitating assembly, as the repetitive sequences are easy to detect, so that tangles in the overlap graph are much easier to solve.

The *Eulerian Superpath Approach* proposed by Pevzner et al. is an excellent approach that can be defined in a mathematically simple way. [76, 74, 75] Although this approach has not yet been applied to the whole-genome shotgun sequencing of large eukaryotes, it is currently the most efficient approach for assembling shotgun reads from a BAC. The biggest difference between the Eulerian Superpath Approach and the other assembly algorithms is that shotgun reads are broken into small pieces and a *de Bruijn graph* is used as the basic structure for computation.

The graph contains the nodes that represent $(k-1)$-mers in the shotgun reads, where k is typically 12-32 bp. For every k-mer in the shotgun reads, an arc from its $(k-1)$-mer prefix to its $(k-1)$-mer suffix is added. Assume that k is 4 and there is a shotgun read ACCGTGACT. The nodes of the graph consist of all 3-mers occurring in the shotgun read, i.e., ACC, CCG, CGT, GTG, TGA, GAC, and ACT. These nodes are connected in the following manner; ACC→CCG, CCG→CGT, CGT→GTG, and so on (Figure 8.42).

If a certain error-correction algorithm completely eliminates the sequencing errors, it can perform the overlap computation implicitly. The average number of outgoing overlaps from shotgun reads in the overlap graph increases quadratically with the sequence coverage. Therefore, the construction of an overlap graph requires quadratic time, whereas it is obvious that a de Bruijn graph can be constructed in linear time. Moreover, it can provide a simpler representation of the ambiguity caused by repetitive sequences. While complicated edges are seen around the repeat sequence in an overlap graph (Figure 8.43), the ambiguity caused by a repeat sequence is represented by two branches in a de Bruijn graph (Figure 8.44).

Although the approach is mathematically simple and seems tolerant of repetitive sequences, recall that it relies on the accuracy of the error-correction algorithm. Virtually all proposed error-correction algorithms that are applicable to mammalian-sized genomes require multiple alignments of reads to detect singleton sequencing errors. Whether the errors in shotgun reads can be corrected without computing multiple alignments, or the equivalent, is still an open problem.

Another problem is the heavy memory requirement. For $k = 20$ or more, the number of nodes in the de Bruijn graph often exceeds two billion when the target is the entire genome of complex eukaryotes. No space-efficient data structure that can fit in memory is known. A breakthrough appears necessary before the Eulerian Path Approach can be applied to whole-genome shotgun sequencing directly.

Although most genome sequencing projects can be classified as "whole-genome shotgun sequencing" or "BAC-by-BAC sequencing," a few cannot. For example, the rat genome sequencing project [22] adopted a hybrid ap-

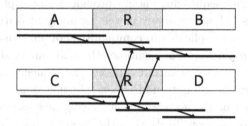

Fig. 8.43 Complicated edges are seen around the repeat.

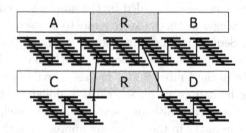

Fig. 8.44 Only two branches are seen at the incoming/outgoing edges of the repeat.

proach that used BAC shotgun reads to localize some of the whole-genome shotgun reads. The question "which is the best approach for whole genome sequencing" remains unanswered.

Assembling a whole genome sometimes poses different algorithmic challenges. For example, the Ciona genome sequencing project [25] suffered from the number of polymorphisms between haplotypes; in some regions of the Ciona genome, as many as 10-15 bp out of 100 bp differed. This proportion far exceeds the sequencing error ratio, making it too difficult to merge two haplotypes. Allowing 15% divergence between shotgun reads leads to significant degradation in the repeat separation capability. Two copies of a retrotransposable element can be separated easily if they have diverged 5% at the DNA level, whereas merging divergent haplotypes means that two divergent copies of a repetitive element might also be merged. Recall that the local sequence coverage of shotgun reads does not always provide a means to detect repetitive regions of the genome.

Another interesting emerging problem is *community shotgun sequencing*, in which microbial communities including unculturable species are shotgun sequenced. Since most microbial species are thought to be unculturable, community shotgun sequencing should provide a global picture of microbial communities. However, many of the assembly algorithms previously reported may not be applicable to community shotgun sequencing because it has unique properties. For example, the shotgun reads may contain many haplotypes from different individuals. The sequence coverage of each species may differ according to its frequency in the community.

As the number of species whose genome sequence has been deciphered increases, it becomes more likely that the target species is evolutionarily close to a species whose genome sequence is available. Assuming that the genome structures are similar, the new assembly may be guided by those of existing species, as described in the previous section. For example, assembly of the chimpanzee genome was guided by the human genome [91].

Note that the Sanger method may give way to new sequencing technologies in the future. Sequencing the human genome was very expensive. What if a 10,000-fold cost reduction were achieved? The National Human Genome Research Institute in the United States has set a goal of reducing the cost of sequencing the human genome to 1,000 U.S. dollars. This may not be achieved in five or ten years, although several new sequencing technologies that may lead the industry in the future are approaching the goal ([17, 62, 94]). These technologies may also drastically change the nature of future assembly problems.

Why do people sequence genomes? One possible answer is "because genomes are there". The problem of genome reconstruction is changing dynamically as new technologies are emerging and more species are becoming targets. Again, we stress that genome sequencing has not yet become routine and remains a complex, dynamic, challenging, and exciting problem.

Problems

Problem 8.1   Vector sequences are sometimes not known because they are prioritized or simply because they have not been sequenced. In such a case, how can we eliminate the vector sequences from the shotgun reads?

Problem 8.2   Eliminating contamination from given shotgun reads is often difficult when moderately significant alignments against the contamination database are found. Long alignments with very high identity occur only when the shotgun reads are contaminated, whereas modest alignments arising from evolutionarily related sequences will be also detected in the contamination elimination step. How should we determine the threshold for the alignment score by which to determine whether a shotgun read includes contamination?

Problem 8.3   Typically, the mitochondrial sequence of the target species is not known. How can we identify the mitochondria sequence within the assembly?

Problem 8.4   Paired pair heuristics do not always connect a pair of reads that truly overlap. In what kind of situation does this process fail?

Problem 8.5   What kind of misassembly does inaccurate estimation of clone length cause? How do the contig lengths affect this?

Problem 8.6   Since the efficiency of cloning varies according to the length of an inserted fragment, the average size of cloned fragments may not reflect the fractionated size directly. How do we identify the size distribution of the actual inserted fragments?

Problem 8.7   What kind of human errors can be involved in the process of whole-genome shotgun sequencing?

Problem 8.8   Part-time workers at a sequencing center have put several incorrect plates in the DNA sequencer, so that incorrect mate pairs are included in the information provided to the assembler, but we do not know which are the incorrect mate pairs. Can we detect them? If so, how?

Problem 8.9 If the target genome has circular plasmids, the read flow graph may contain cycles. How should we modify the assembly algorithm?

Problem 8.10 Unlike short plasmids, fosmids/cosmids have the special property that they cannot accommodate fragments that are too short due to biological constraints. How should this property be utilized?

Problem 8.11 Tandem repeats, such as telomere and centromere repeats, are often not reconstructed in the whole-genome shotgun assembly. Why?

Problem 8.12 How do greedy algorithms assemble a 1,300-bp tandem repeat sequence in which the unit repeat consists of 100 bp? Assume that the read length is exactly 500 bp if necessary.

Problem 8.13 If there are recently duplicated segments in the target genome, what happens to the whole-genome shotgun assembly?

Problem 8.14 Assume that the polymorphism rate between two haplotype of a target species is 1%. What should the assembly algorithm do?

Problem 8.15 Assume that the polymorphism rate between the two haplotypes of a target species is 10%. What should the assembly algorithm do? How should we modify the design of the shotgun libraries?

Problem 8.16 If the DNA source is from a male (female), the shotgun read set contains shotgun reads from X/Y (Z/W) chromosomes and they appear with half coverage as compared to those from autosomes. How should we treat them?

Problem 8.17 If one decides to merge two haplotypes that have 3% polymorphisms, how can one distinguish repetitive sequences from the divergent haplotypes?

Problem 8.18 Short plasmids are rarely chimeric. The ratio is usually less than 0.1%. However, the mate pairs between the contigs are more likely to be chimeric. Why?

Problem 8.19 A mate pair that consists of a repetitive shotgun read and a unique shotgun read may resolve tangles caused by repetitive sequences. In what kind of situation does this occur? An algorithm to resolve the situation seems obvious; however, most existing assemblers do not incorporate the algorithm. Why?

Problem 8.20 Longer clones are more expensive to sequence, while short clones are less expensive to sequence. How should we design the shotgun libraries?

Software Availability

The Java source codes described are available from the following Web site for further study.

> http://mlab.cb.k.u-tokyo.ac.jp/~moris/LSGSP/

These codes can be used under the conditions that the site does not form part of this book, nor is it maintained by the publisher; there are no guarantees regarding the reliability of the source codes.

Bibliography

[1] Mohamed Ibrahim Abouelhoda and Enno Ohlebusch. Chaining algorithms for multiple genome comparison. *Journal of Discrete Algorithms*, 3:321–341, 2005. 137

[2] Stephen F. Altschul, Warren Gish, Webb Miller, Eugene W. Myers, and David J. Lipman. Basic local alignment search tool. *J. Mol. Biol.*, 215(3):403–410, 1990. 131

[3] Stephen F. Altschul, Thomas L. Madden, Alejandro A. Schaffer, Jinghui Zhang, Zheng Zhang, Webb Miller, and David J. Lipman. Gapped BLAST and PSI-BLAST: a new generation of protein database search programs. *Nucleic Acids Research*, 25(17):3389–402, 1997. 131

[4] Eric L. Anson and Eugene W. Myers. Realigner: a program for refining DNA sequence multi-alignments. *Proceedings of the first annual international conference on Computational molecular biology 9-16 ACM Press, Santa Fe, New Mexico, United*, page States, 1997. 210

[5] Samuel Aparicio, Jarrod Chapman, Elia Stupka, Nik Putnam, Jer ming Chia, Paramvir Dehal, Alan Christoffels, Sam Rash, Shawn Hoon, Arian Smit, Maarten D. Sollewijn Gelpke, Jared Roach, Tania Oh, Isaac Y. Ho, Marie Wong, Chris Detter, Frans Verhoef, Paul Predki, Alice Tay, Susan Lucas, Paul Richardson, Sarah F. Smith, Melody S. Clark, Yvonne J. K. Edwards, Norman Doggett, Andrey Zharkikh, Sean V. Tavtigian, Dmitry Pruss, Mary Barnstead, Cheryl Evans, Holly Baden, Justin Powell, Gustavo Glusman, Lee Rowen, Leroy Hood, Y. H. Tan, Greg Elgar, Trevor Hawkins, Byrappa Venkatesh, Daniel Rokhsar, and Sydney Brenner. Whole-genome shotgun assembly and analysis of the genome of Fugu rubripes. *Science*, 297:1301–1310, 2002. 194, 195, 208

[6] Ricardo A. Baeza-Yates and Chris H. Perleberg:. Fast and practical approximate string matching. In *Proceedings of Combinatorial Pattern Matching, Third Annual Symposium, CPM 92*, pages 185–192, 1992. 146

[7] Serafim Batzoglou, David B. Jaffe, Ken Stanley, Jonathan Butler, Sante Gnerre, Evan Mauceli, Bonnie Berger, Jill P. Mesirov, and Eric S. Lander. ARACHNE: a whole-genome shotgun assembler. *Genome Res*, 12:177–189, 2002. 195

[8] Jon Louis Bentley and M. Douglas McIlroy. Engineering a sort function. *j-SPE*, 23(11):1249–1265, November 1993. **32**

[9] Robert W. Blakesley, Nancy F. Hansen, James C. Mullikin, Pamela J. Thomas, Jennifer C. McDowell, Baishali Maskeri, Alice C. Young, Beatrice Benjamin, Shelise Y. Brooks, Bradley I. Coleman, Jyoti Gupta, Shi-Ling Ho, Eric M. Karlins, Quino L. Maduro, Sirintorn Stantripop, Cyrus Tsurgeon, Jennifer L. Vogt, Michelle A. Walker, Catherine A. Masiello, Xiaobin Guan, NISC Comparative Sequencing Program, Gerard G. Bouffard, and Eric D. Green. An intermediate grade of finished genomic sequence suitable for comparative analyses. *Genome Res*, 14:2235–2244, 2004. **163**

[10] Robert S. Boyer and J. Strother Moore. A fast string search algorithm. *Communications of the ACM*, 20(10):762–772, 1977. **94**

[11] Daniel G. Brown, Ming Li, and Bin Ma. A tutorial of recent developments in the seeding of local alignment. *J. of Bioinformatics and Comp. Biol.*, 2(4):819–842, December 2004. **134**

[12] Jeremy Buhler, Uri Keich, and Yanni Sun. Designing seeds for similarity search in genomic DNA. In *Proceedings of the Annual International Conference on Computational Molecular Biology (RECOMB-2003)*, pages 67–75, 2003. **126, 134**

[13] Stefan Burkhardt, Andreas Crauser, Paolo Ferragina, Hans-Peter Lenhof, Eric Rivals, and Martin Vingron. q-gram based database searching using a suffix array. In *Proceedings of the 3rd Annual International Conference on Computational Molecular Biology (RECOMB-99)*, 1999. **131**

[14] Stefan Burkhardt and Juha Karkkainen. Better filtering with gapped q-grams. *FUNDINF: Fundamenta Informatica*, 56, 2003. **134**

[15] Stefan Burkhardt and Juha Kärkkäinen. Fast lightweight suffix array construction and checking. In *CPM*, pages 55–69, 2003. **81**

[16] Andrea Califano and Isidore Rigoutsos. FLASH: A fast look-up algorithm for string homology. In *ISMB*, pages 56–64, 1993. **134**

[17] Eugene Y. Chan. Advances in sequencing technology. *Mutat Res*, 573:13–40, 1996. **216**

[18] Kwok Pui Choi, Fanfan Zeng, and Louxin Zhang. Good spaced seeds for homology search. *Bioinformatics*, 20(7):1053–1059, 2004. **134**

[19] Kwok Pui Choi and Louxin Zhang. Sensitivity analysis and efficient method for identifying optimal spaced seeds. *JCSS: Journal of Computer and System Sciences*, 68, 2004. **134**

[20] International Human Genome Sequencing Consortium. Initial sequencing and analysis of the human genome. *Nature*, 409:860–921, 2001. **194**

[21] Mouse Genome Sequencing Consortium. Initial sequencing and comparative analysis of the mouse genome. *Nature*, 420:520–562, 2002. **172, 208**

[22] Rat Genome Sequencing Project Consortium. Genome sequence of the brown norway rat yields insights into mammalian evolution. *Nature*, 428:493–521, 2004. **187, 208, 215**

[23] Thomas H. Cormen, Charles E. Leiserson, and Ronald L. Rivest. *Introduction to Algorithms*. MIT Press, 1990. **44, 97, 138**

[24] Frank B. Dean, John R. Nelson, Theresa L. Giesler, and Roger S. Lasken.

Rapid amplification of plasmid and phage DNA using phi 29 DNA polymerase and multiply-primed rolling circle amplification. *Genome Res*, 11:1095–1099, 2001. 163

[25] Paramvir Dehal, Yutaka Satou, Robert K. Campbell, Jarrod Chapman, Bernard Degnan, Anthony De Tomaso, Brad Davidson, Anna Di Gregorio, Maarten Gelpke, David M. Goodstein, Naoe Harafuji, Kenneth E. M. Hastings, Isaac Ho, Kohji Hotta, Wayne Huang, Takeshi Kawashima, Patrick Lemaire, Diego Martinez, Ian A. Meinertzhagen, Simona Necula, Masaru Nonaka, Nik Putnam, Sam Rash, Hidetoshi Saiga, Masanobu Satake, Astrid Terry, Lixy Yamada, Hong-Gang Wang, Satoko Awazu, Kaoru Azumi, Jeffrey Boore, Margherita Branno, Stephen Chin-bow, Rosaria DeSantis, Sharon Doyle, Pilar Francino, David N. Keys, Shinobu Haga, Hiroko Hayashi, Kyosuke Hino, Kaoru S. Imai, Kazuo Inaba, Shungo Kano, Kenji Kobayashi, Mari Kobayashi, Byung-In Lee, Kazuhiro W. Makabe, Chitra Manohar, Giorgio Matassi, Monica Medina, Yasuaki Mochizuki, Steve Mount, Tomomi Morishita, Sachiko Miura, Akie Nakayama, Satoko Nishizaka, Hisayo Nomoto, Fumiko Ohta, Kazuko Oishi, Isidore Rigoutsos, Masako Sano, Akane Sasaki, Yasunori Sasakura, Eiichi Shoguchi, Tadasu Shin-i, Antoinetta Spagnuolo, Didier Stainier, Miho M. Suzuki, Olivier Tassy, Naohito Takatori, Miki Tokuoka, Kasumi Yagi, Fumiko Yoshizaki, Shuichi Wada, Cindy Zhang, P. Douglas Hyatt, Frank Larimer, Chris Detter, Norman Doggett, Tijana Glavina, Trevor Hawkins, Paul Richardson, Susan Lucas, Yuji Kohara, Michael Levine, Nori Satoh, and Daniel S. Rokhsar. The draft genome of Ciona intestinalis: insights into chordate and vertebrate origins. *Science*, 298:2157–2167, 2002. 208, 216

[26] Arthur L. Delcher, Adam Phillippy, Jane Carlton, and Steven L. Salzberg. Fast algorithms for large-scale genome alignment and comparison. *Nucleic Acids Research*, 30(11):2478–2483, 2002. 131

[27] Koichiro Doi, Tomoyuki Yamada, and Shinich Morishita. Elastically-spaced seeds – an extension to spaced seeds. *Unpublished manuscript*, 2005. 133

[28] David Eppstein, Zvi Galil, Raffaele Giancarlo, and Giuseppe F. Italiano. Sparse dynamic programming I: linear cost functions. *J. ACM*, 39(3):519–545, 1992. 137

[29] A. Fire and S.Q. Xu. Rolling replication of short DNA circles. *Proc Natl Acad Sci USA*, 92:4641–4645, 1995. 163

[30] Liliana Florea, George Hartzell, Zheng Zhang, Gerald M. Rubin, and Webb Miller. A computer program for aligning a cDNA sequence with a genomic DNA sequence. *Genome Research*, 8(9):967–74, 1998. 131

[31] Osamu Gotoh. An improved algorithm for matching biological sequences. *J. Mol. Biol.*, 162:705–708, 1982. 117

[32] P. Green. phrap web site. http://www.phrap.org/. 164, 184, 185

[33] Phil Green. Whole-genome disassembly. *Proc Natl Acad Sci USA*, 99:4143–4144, 2002. 213

[34] Philip Green. Against a whole-genome shotgun. *Genome Res*, 7:410–417, 1997. 213

[35] Dan Gusfield. *Algorithms on Strings, Trees, and Sequences*. Cambridge Uni-

versity Press, 1997. 53, 97

[36] Daniel S. Hirschberg. A linear space algorithm for computing maximal common subsequences. *Commun. ACM*, 18(6):341–343, 1975. 120

[37] Charles Antony Richard Hoare. Quicksort. *Comput. J*, 5(1):10–15, 1962. 22

[38] Xiaoqiu Huang, Jianmin Wang, Srinivas Aluru, Shiaw-Pyng Yang, and LaDeana Hillier. PCAP: a whole-genome assembly program. *Genome Res*, 13:2164–2170, 2003. 185, 194, 195

[39] David B. Jaffe, Jonathan Butler, Sante Gnerre, Evan Mauceli, Kerstin Lindblad-Toh, Jill P. Mesirov, Michael C. Zody, and Eric S. Lander. Whole-genome sequence assembly for mammalian genomes: ARACHNE 2. *Genome Res*, 13:91–96, 2003. 184, 210

[40] O. Jaillon, J.M. Aury, F. Brunet, J.L. Petit, N. Stange-Thomann, E. Mauceli, L. Bouneau, C. Fischer, C. Ozouf-Costaz, A. Bernot, S. Nicaud, D. Jaffe, S. Fisher, G. Lutfalla, C. Dossat, B. Segurens, C. Dasilva, M. Salanoubat, M. Levy, N. Boudet, S. Castellano, V. Anthouard, C. Jubin, V. Castelli, M. Katinka, B. Vacherie, C. Biemont, Z. Skalli, L. Cattolico, J. Poulain, V. De Berardinis, C. Cruaud, S. Duprat, P. Brottier, J.P. Coutanceau, J. Gouzy, G. Parra, G. Lardier, C. Chapple, K.J. McKernan, P. McEwan, S. Bosak, M. Kellis, J.N. Volff, R. Guigo, M.C. Zody, J. Mesirov, K. Lindblad-Toh, B. Birren, C. Nusbaum, D. Kahn, M. Robinson-Rechavi, V. Laudet, V. Schachter, F. Quetier, W. Saurin, C. Scarpelli, P. Wincker, E.S. Lander, J. Weissenbach, and H. Roest Crollius. Genome duplication in the teleost fish tetraodon nigroviridis reveals the early vertebrate proto-karyotype. *Nature*, 431:946–957, 2004. 208

[41] Deborah Joseph, Joao Meidanis, and Prasoon Tiwari. Determining DNA sequence similarity using maximum independent set algorithms for interval graphs. In *SWAT*, pages 326–337, 1992. 137

[42] Juha Karkkainen and Peter Sanders. Simple linear work suffix array construction. In *ICALP: Annual International Colloquium on Automata, Languages and Programming*, 2003. 74, 75

[43] Richard M. Karp, Raymond E. Miller, and Arnold L. Rosenberg. Rapid identification of repeated patterns in strings, trees, and arrays. In *STOC: ACM Symposium on Theory of Computing (STOC)*, 1972. 68

[44] Richard M. Karp and Michael O. Rabin. Efficient randomized pattern-matching algorithms. Technical report, Aiken Computation Laboratory, Harvard University, 1981. 85

[45] Toru Kasai, Gunho Lee, Hiroki Arimura, Setsuo Arikawa, and Kunsoo Park. Linear-time longest-common-prefix computation in suffix arrays and its applications. In *CPM: 12th Symposium on Combinatorial Pattern Matching*, 2001. 62

[46] John D. Kececioglu and Eugene W. Myers. Combinatorial algorithms for DNA sequence assembly. *Algorithmica*, 13(1/2):7–51, 1995. 175

[47] W. James Kent. BLAT–the BLAST-like alignment tool. *Genome Research*, 12(4):656–642, 2002. 131, 134

[48] Donald E. Knuth. *The Art of Computer Programming, Volume 3: Sorting*

and Searching. Addison-Wesley, Reading, MA., 1978. 13, 28, 37, 44
[49] Donald E. Knuth, James H. Morris, and Vaughan R. Pratt. Fast pattern matching in strings. SICOMP: SIAM Journal on Computing, 6, 1977. 88
[50] Pang Ko and Srinivas Aluru. Space efficient linear time construction of suffix arrays. In CPM: 14th Symposium on Combinatorial Pattern Matching, 2003. 74
[51] A. Koga, M. Suzuki, H. Inagaki, Y. Bessho, and H. Hori. Transposable element in fish. Nature, 383:30, 1996. 198
[52] Stefan Kurtz. Reducing the space requirement of suffix trees. Software - Practice and Experience, 29(13):1149–1171, 1999. 53
[53] Eric S. Lander and Michael S. Waterman. Genomic mapping by fingerprinting random clones: a mathematical analysis. Genomics, 2(3):231–239, 1988. 170
[54] N. Jesper Larsson and Kunihiko Sadakane. Faster suffix sorting. Technical report, Lund University, 1999. 69
[55] Ming Li, Bin Ma, Derek Kisman, and John Tromp. PatternHunter II: Highly sensitive and fast homology search. J. of Bioinformatics and Comp. Biol., 2(3):417–439, September 2004. 134
[56] Wen-Hsiung Li. Molecular Evolution. Sunderland, Mass., USA, 1997. 155
[57] Paul M. Lizardi, Xiaohua Huang, Zhengrong Zhu, Patricia Bray-Ward, David C. Thomas, and David C. Ward. Mutation detection and single-molecule counting using isothermal rolling-circle amplification. Nat Genet, 19:225–232, 1998. 163
[58] Bin Ma, John Tromp, and Ming Li. PatternHunter faster and more sensitive homology search. Bioinformatics, 18(3):440–445, 2002. 134
[59] Udi Manber and Gene Myers. Suffix arrays: A new method for on-line string searches. In Proceedings of the First Annual ACM-SIAM Symposium on Discrete Algorithms, pages 319–327, San Francisco, California, 22–24 January 1990. 54, 69
[60] Udi Manber and Gene Myers. Suffix arrays: A new method for on-line string searches. SICOMP: SIAM Journal on Computing, 22, 1993. 54, 69
[61] Giovanni Manzini and Paolo Ferragina. Engineering a lightweight suffix array construction algorithm. Algorithmica, 40(1):33–50, 2004. 81
[62] Marcel Margulies, Michael Egholm, William E. Altman, Said Attiya, Joel S. Bader, Lisa A. Bemben, Jan Berka, Michael S. Braverman, Yi-Ju Chen, Zhoutao Chen, Scott B. Dewell, Lei Du, Joseph M. Fierro, Xavier V. Gomes, Brian C. Godwin, Wen He, Scott Helgesen, Chun He Ho, Gerard P. Irzyk, Szilveszter C. Jando, Maria L. I. Alenquer, Thomas P. Jarvie, Kshama B. Jirage, Jong-Bum Kim, James R. Knight, Janna R. Lanza, John H. Leamon, Steven M. Lefkowitz, Ming Lei, Jing Li, Kenton L. Lohman, Hong Lu, Vinod B. Makhijani, Keith E. McDade, Michael P. McKenna, Eugene W. Myers, Elizabeth Nickerson, John R. Nobile, Ramona Plant, Bernard P. Puc, Michael T. Ronan, George T. Roth, Gary J. Sarkis, Jan Fredrik Simons, John W. Simpson, Maithreyan Srinivasan, Karrie R. Tartaro, Alexander Tomasz, Kari A. Vogt, Greg A. Volkmer, Shally H. Wang, Yong Wang, Michael P. Weiner, Pengguang Yu, Richard F. Begley,

and Jonathan M. Rothberg. Genome sequencing in microfabricated high-density picolitre reactors. *Nature*, 437:376–380, 2005. 216

[63] Edward M. McCreight. A space-economical suffix tree construction algorithm. *Journal of the ACM*, 23(2):262–272, April 1976. 53

[64] James C. Mullikin and Zemin Ning. The PHUSION assembler. *Genome Res*, 13:81–90, 2003. 195, 204

[65] David R. Musser. Introspective sorting and selection algorithms. *Software - Practice and Experience*, 27(8):983–993, 1997. 31

[66] Eugene W. Myers. Toward simplifying and accurately formulating fragment assembly. *J Comput Biol*, 2:275–290, 1995. 175, 196

[67] Eugene W. Myers, Granger G. Sutton, Art L. Delcher, Ian M. Dew, Dan P. Fasulo, Michael J. Flanigan, Saul A. Kravitz, Clark M. Mobarry, Knut H. J. Reinert, Karin A. Remington, Eric L. Anson, Randall A. Bolanos, Hui-Hsien Chou, Catherine M. Jordan, Aaron L. Halpern, Stefano Lonardi, Ellen M. Beasley, Rhonda C. Brandon, Lin Chen, Patrick J. Dunn, Zhongwu Lai, Yong Liang, Deborah R. Nusskern, Ming Zhan, Qing Zhang, Xiangqun Zheng, Gerald M. Rubin, Mark D. Adams, and J. Craig Venter. A whole-genome assembly of Drosophila. *Science*, 287:2196–2204, 2000. 196, 210

[68] Gene Myers and Webb Miller. Chaining multiple-alignment fragments in sub-quadratic time. In *SODA*, pages 38–47, 1995. 137

[69] Yuki Naito, Tomoyuki Yamada, Kumiko Ui-Tei, Shinichi Morishita, and Kaoru Saigo. sidirect: highly effective, target-specific siRNA design software for mammalian RNA interference. *Nucleic Acids Research*, 32(Web-Server-Issue):124–129, 2004. 144

[70] Saul B. Needleman and Christian D. Wunsch. A general method applicable to the search for similarities in the amino acid sequence of two proteins. *J. Mol. Biol.*, pages 443–453, 1970. 105

[71] Zemin Ning, Anthony J. Cox, and James C. Mullikin. SSAHA: a fast search method for large DNA databases. *Genome Research*, 11(10):1725–9, 2001. 131

[72] Jun Ogasawara and Shinichi Morishita. A fast and sensitive algorithm for aligning ESTs to the human genome. *J. Bioinform. Comput. Biol.*, 1(2):363–86, 2003. 129

[73] Hannu Peltola, Hans Soderlund, and Esko Ukkonen. Seqaid: a DNA sequence assembling program based on a mathematical model. *Nucleic Acids Res*, 12:307–321, 1984. 175

[74] Pavel A. Pevzner and Haixu Tang. Fragment assembly with double-barreled data. *Bioinformatics 17 Suppl*, 1:S225–S233, 2001. 214

[75] Pavel A. Pevzner, Haixu Tang, and Michael S. Waterman. An eulerian path approach to DNA fragment assembly. *Proc Natl Acad Sci USA*, 98:9748–9753, 2001. 214

[76] Pavel A. Pevzner, Haixu Tang, and Michael S. Waterman. A new approach to fragment assembly in DNA sequencing. *RECOMB 2001,*, pages 256–265, 2001. 214

[77] Pavel A. Pevzner and Michael S. Waterman. Multiple filtration and approx-

imate pattern matching. *Algorithmica*, 13(1/2):135–154, January/February 1995. 134

[78] Mihai Pop, Adam Phillippy, Arthur L. Delcher, and Steven L. Salzberg. Comparative genome assembly. *Brief Bioinform*, 5:237–248, 2004. 170

[79] Simon J. Puglisi, William F. Smyth, and Andrew Turpin. The performance of linear time suffix sorting algorithms. In *Proceedings of Data Compression Conference (DCC'05)*, pages 358–367, 2005. 79

[80] Simon J. Puglisi, William F. Smyth, and Andrew Turpin. A taxonomy of suffix array construction algorithms. In *Proceedings of the Prague Stringology Conference (PSC'05)*, pages 1–30, 2005. 81

[81] Sven Rahmann. Fast large scale oligonucleotide selection using the longest common factor approach. *J. of Bioinformatics and Comp. Biol.*, 1(2):343–61, 2003. 143

[82] Michael J. Reagin, Theresa L. Giesler, Alia L. Merla, Jeanine M. Resetar-Gerke, Kinga M. Kapolka, and J. Anthony Mamone. Templiphi: A sequencing template preparation procedure that eliminates overnight cultures and DNA purification. *J Biomol Tech*, 14:143–148, 2003. 163

[83] S. Richards, Y. Liu, B.R. Bettencourt, P. Hradecky, S. Letovsky, R. Nielsen, K. Thornton, M.J. Hubisz, R. Chen, R.P. Meisel, O. Couronne, S. Hua, M.A. Smith, P. Zhang, J. Liu, H.J. Bussemaker, M.F. van Batenburg, S.L. Howells, S.E. Scherer, E. Sodergren, B.B. Matthews, M.A. Crosby, A.J. Schroeder, D. Ortiz-Barrientos, C.M. Rives, M.L. Metzker, D.M. Muzny, G. Scott, D. Steffen, D.A. Wheeler, K.C. Worley, P. Havlak, K.J. Durbin, A. Egan, R. Gill, J. Hume, M.B. Morgan, G. Miner, C. Hamilton, Y. Huang, L. Waldron, D. Verduzco, K.P. Clerc-Blankenburg, I. Dubchak, M.A. Noor, W. Anderson, K.P. White, A.G. Clark, S.W. Schaeffer, W. Gelbart, G.M. Weinstock, and R.A. Gibbs. Comparative genome sequencing of Drosophila pseudoobscura: chromosomal, gene, and cis-element evolution. *Genome Res*, 15:1–18, 2005. 208

[84] M. Roberts, W. Hayes, B.R. Hunt, S.M. Mount, and J.A. Yorke. Reducing storage requirements for biological sequence comparison. *Bioinformatics*, 20:3363–3369, 2004. 191

[85] Meena Kishore Sakharkar, Vincent T.K. Chow, and Pandjassarame Kangueane. Distributions of exons and introns in the human genome. *In Silico Biology*, Vol. 4, 2004. 114

[86] F. Sanger, A. R. Coulson, B. G. Barrell, A. J. Smith, and B. A. Roe. Cloning in single-stranded bacteriophage as an aid to rapid DNA sequencing. *J Mol Biol*, 143(2):161–178, 1980. 155, 171

[87] F. Sanger, A.R. Coulson, G.F. Hong, D.F. Hill, and G.B. Petersen. Nucleotide sequence of bacteriophage lambda DNA. *J Mol Biol*, 162:729–773, 1982. 157

[88] Arnold Schonhage, Mike Paterson, and Nicholas Pippenger. Finding the median. *Journal of Computer and System Sciences*, 13(2):184–199, 1976. 74

[89] D.C. Schwartz, X. Li, L.I. Hernandez, S.P. Ramnarain, E.J. Huff, and Y.K. Wang. Ordered restriction maps of saccharomyces cerevisiae chromosomes

constructed by optical mapping. *Science*, 262:110–114, 1993. 169
[90] Scott Schwartz, W. James Kent, Arian Smit, Zheng Zhang, Robert Baertsch, Ross C. Hardison, David Haussler, and Webb Miller. Human-mouse alignments with BLASTZ. *Genome Research*, 13(1):103–7, 2003. 134
[91] The Chimpanzee Sequencing and Analysis Consortium. Initial sequence of the chimpanzee genome and comparison with the human genome. *Nature*, 437:69–87, 2005. 136, 216
[92] Julian Seward. On the performance of BWT sorting algorithms. In *Data Compression Conference*, pages 173–182, 2000. 81
[93] Harris Shapiro. Outline of the assembly process: JAZZ, the JGI in-house assembler. *Lawrence Berkeley National Laboratory Paper*, pages LBNL-58236, 2005. 194, 195
[94] Jay Shendure, Gregory J. Porreca, Nikos B. Reppas, Xiaoxia Lin, John P. McCutcheon, Abraham M. Rosenbaum, Michael D. Wang, Kun Zhang, Robi D. Mitra, and George M. Church. Accurate multiplex polony sequencing of an evolved bacterial genome. *Science*, 309:1728–1732, 2005. 216
[95] Tetsuo Shibuya and Igor Kurochkin. Match chaining algorithms for cDNA mapping. In *WABI*, pages 462–475, 2003. 137
[96] L.M. Smith, J.Z. Sanders, R.J. Kaiser, P. Hughes, C. Dodd, C.R. Connell, C. Heiner, S.B. Kent, and L.E. Hood. Fluorescence detection in automated DNA sequence analysis. *Nature*, 321:674–679, 1986. 159
[97] Temple F. Smith and Michael S. Waterman. Identification of common molecular subsequences. *Journal of Molecular Biology*, Vol. 147:195–197, 1981. 108
[98] Wing-Kin Sung and Wah-Heng Lee. Fast and accurate probe selection algorithm for large genomes. In *CSB*, pages 65–74. IEEE Computer Society, 2003. 147
[99] S. Tabor and C.C. Richardson. DNA sequence analysis with a modified bacteriophage T7 DNA polymerase. *Proc Natl Acad Sci USA*, 84:4767–4771, 1987. 160
[100] S. Tabor and C.C. Richardson. A single residue in DNA polymerases of the Escherichia coli DNA polymerase I family is critical for distinguishing between deoxy- and dideoxyribonucleotides. *Proc Natl Acad Sci USA*, 92:6339–6343, 1995. 160
[101] J.D. Thompson, D.G. Higgins, and T.J. Gibson. ClustalW: improving the sensitivity of progressive multiple sequence alignment through sequence weighting, position-specific gap penalties and weight matrix choice. *Nucleic Acids Res*, 22:4673–4680, 1994. 181
[102] Esko Ukkonen. On-line construction of suffix trees. *Algorithmica*, 14(3):249–260, September 1995. 53
[103] Jun Wang, Gane Ka-Shu Wong, Peixiang Ni, Yujun Han, Xiangang Huang, Jianguo Zhang, Chen Ye, Yong Zhang, Jianfei Hu, Kunlin Zhang, Xin Xu, Lijuan Cong, Hong Lu, Xide Ren, Xiaoyu Ren, Jun He, Lin Tao, Douglas A. Passey, Jian Wang, Huanming Yang, Jun Yu, and Songgang Li. RePS: a sequence assembler that masks exact repeats identified from the shotgun data. *Genome Res*, 12:824–831, 2002. 195

[104] Peter Weiner. Linear pattern matching algorithms. In *FOCS*, pages 1–11, 1973. 52

[105] Tomoyuki Yamada and Shinichi Morishita. Computing highly specific and noise tolerant oligomers efficiently. *J. of Bioinformatics and Comp. Biol.*, 2(1):21–46, 2004. 141

[106] Tomoyuki Yamada and Shinichi Morishita. Accelerated off-target search algorithm for siRNA. *Bioinformatics*, 21(8):1316–1324, 2005. 150, 151, 153

[107] Zheng Zhang, Scott Schwartz, Lukas Wagner, and Webb Miller. A greedy algorithm for aligning DNA sequences. *J. Comput. Biol.*, 7(1-2):203–14, 2000. 131

Index

h-group, 70
h-ordering, 68
h-rank, 70
lcf, 67
lcp, 58
-, 99
3'-exonuclease activity, 157
5-bromo-4-chloro-3-indoyl-B-D-galactoside, 163

affine gap penalty, 114
alignment path, 103
alignments, 99
alignment score, 103
approximate inverse suffix array, 70
approximate string search, 99
approximate suffix array, 68
ARACHNE, 184, 190, 195, 201, 210
assembler, 156
asymptotic upper bound, 16

BAC, 165
BAC-by-BAC, 213
bacterial artificial chromosome, 165
bacteriophage lambda, 155
bad character heuristics, 94
banded alignment, 140
base calling, 164
Bernoulli random distribution, 126
binary search, 7
binary search of suffix array, 56

BLAST, 131
BLASTZ, 134
BLAT, 134
Boyer-Moore algorithm, 94
breaking and rejoining, 210
brute-force search, 2
brute-force string search, 86
BYP method, 146

cDNA sequences, 114
chain, 137
chaining, 135
chimeric clone, 162, 205
chimeric read, 204
clone coverage, 207
cloning, 161
common factor, 67
community shotgun sequencing, 216
comparative genomics, 135
consensus, 175, 180, 181, 210
consensus sequence, 181, 210
contamination removal, 187
contig, 166
cosmid, 165

ddATP, 157
ddCTP, 157
ddGTP, 157
ddNTPs, 157
ddTTP, 157
de Bruijn graph, 214
deletion, 101

dideoxyadenosine triphosphate, 157
dideoxycytosine triphosphate, 157
dideoxyguanosine triphosphate, 157
dideoxy method, 157
dideoxytyrosine triphosphate, 157
digits, 33
direct-address table, 39
direct-address table size, 45
divide-and-conquer approach, 12
DNAmicroarray, 141
dNTPs, 157
double-barreled shotgun sequencing, 168
double-stranded assembly, 172
double-stranded reads, 173
doubling technique, 68
dynamic programming approach, 105

edit graph, 103
edit operations, 101
EDVAC computer, 13
electropherogram, 163
electrophoresis, 158
error correction, 199
Eulerian Superpath Approach, 214
exons, 114

FISH, 169
fluorescence insitu hybridization, 169
footprint, 120
fosmid, 165
four-color fluorescent dye method, 159

gap, 99
Gapped Blast, 131
gapped seeds, 134
gap penalty, 103
genetic map, 169
global alignment, 101
global alignment path, 103
Gotoh's algorithm, 116
greedy merging approach, 191
greedy scaffolding, 206

Hamming distance, 143

hash function, 41
hash table, 41
hash table size, 45
heap, 17
heap property, 17
heap sort, 17
Hirschberg's space reduction technique, 120

insertion, 101
insertion sort, 11
introns, 114
introspective sort, 31
inverse suffix array, 54
iterative improvement, 193
iterative improvements, 208

JAZZ, 194, 195

k-mer(string), 4
k-mer integer, 4
Kärkkäinen-Sanders algorithm, 75
KMP shift rule, 89
Knuth-Morris-Pratt algorithm, 88

LacZ, 162
Lander-Waterman statistics, 170
Larsson-Sadakane algorithm, 69
layout, 175, 177
least significant digit, 33
level of anode, 17
LINE, 202
linear gap penalty criterion, 114
local alignment, 101
local alignment path, 103
local alignment paths, 108
longest common factor, 67, 141
longest common prefix, 58
longest suffix-prefix match, 90
long interspersed nucleotide elements, 202

Master-Slave alignment, 181
match ratio, 127
match score, 102
mate pair, 166

mate pair constraints, 213
maximal read heuristics, 201
MCS, 162
Mega Blast, 131
merge sort, 12
Mersenne prime number, 84
minimizer, 191
mismatch penalty, 102
mismatch tolerance, 143
most significant digit, 33
multicloning site, 162
multiple hits of seeds, 131
MUMmer 2, 131

N50 contig length, 211
N50 scaffold length, 211
N50 value, 211
Needleman-Wunsch algorithm, 105
nonoverlapping substrings, 48

occurrence frequency, 65, 141
one-origin indexing, 1
optical map, 169
optical mapping, 169
overlap, 175
Overlap-Layout-Consensus, 175
overlap alignment, 111
overlap alignment path, 111
overlap graph, 176
overlapping, partially matching seeds, 151

paired pair, 202
partially matching seeds, 147
PCR, 168
PCR primer, 141
phred, 164
PHUSION, 195, 204
pivot, 23, 32
plasmid, 165
Polymerase Chain Reaction, 168
pre-consensus, 181
predecessor, 137
prefix, 51
proper prefix, 51
proper suffix, 51

pUC18, 162, 165

quality trimming, 186
quality values, 164
QUASAR, 131
QV, 164

Rabin-Karp algorithm, 83
Radiation Hybrid, 169
radix sort, 33
RAMEN, 184
randomized quick sort, 22
read, 166, 168
read flow graph, 191
redundant seed search trials, 149
repeat boundary, 198
repeat database, 196
repeat separation strategy, 200
RePS, 195
RH, 169

Sanger, 155, 157
Sanger method, 156, 157
scaffold, 168, 180
scaffolding, 180
score, 103
seed and extend strategy, 125, 188
seeded alignment, 125
seed hits, 145
seeds, 188
seed search trials, 145
sensitivity of seeds, 127
sequence coverage, 171
sequence cover ratio, 171
sequence tagged site, 169
short interspersed nuclear elements, 194
shotgun sequencing, 168
sim4, 131
SINE, 194, 202
sister clones, 185
sister edge, 202
Smith-Waterman algorithm, 108
specificity, 127
SSAHA, 131
stack, 29

stack depth, 29
startup gap penalty, 114
stochastic gaps, 166
STS, 169
substitution, 101
successor, 137
suffix, 51
suffix-prefix match, 88
suffix-prefix matches, 111
suffix array, 54
suffix trees, 52
super contig, 168
synteny information, 170

tail recursion, 29
tail recursion elimination, 29

ternary split quick sort, 32
trace back procedure, 108, 109, 120

untangling, 179

vector masking, 183

WGS, 155
whole genome shotgun, 155
window-trimming approach, 186
worst-case execution time, 16

X-gal, 163

zero-origin indexing, 1